AF798460

Sustainable Urban Futures

Series Editor
Sina Shahab
School of Geography and Planning
Cardiff University
Cardiff, UK

This series includes a broad range of Pivot length books offering accessible and applied texts designed to appeal to both practitioners and academics in the field. Pivots in the series will explore how sustainability can be achieved in Future Cities and how technology can assist in supporting sustainable transitions to better respond to the urgencies of climate change, equity needs and inclusivity aligning the two core themes of Urban Science and Future Science.

Antonio José Salvador

Metropolitan Food Systems

European Trajectories in Planning, Governance and Landscape Management

palgrave
macmillan

Antonio José Salvador
Dipartimento di Architettura e Studi Urbani
Politecnico di Milano
MILANO, Italy

ISSN 2730-6607 ISSN 2730-6615 (electronic)
Sustainable Urban Futures
ISBN 978-3-032-26015-4 ISBN 978-3-032-26016-1 (eBook)
https://doi.org/10.1007/978-3-032-26016-1

© The Editor(s) (if applicable) and The Author(s), under exclusive license to Springer Nature Switzerland AG 2026

This work is subject to copyright. All rights are solely and exclusively licensed by the Publisher, whether the whole or part of the material is concerned, specifically the rights of translation, reprinting, reuse of illustrations, recitation, broadcasting, reproduction on microfilms or in any other physical way, and transmission or information storage and retrieval, electronic adaptation, computer software, or by similar or dissimilar methodology now known or hereafter developed.
The use of general descriptive names, registered names, trademarks, service marks, etc. in this publication does not imply, even in the absence of a specific statement, that such names are exempt from the relevant protective laws and regulations and therefore free for general use.
The publisher, the authors and the editors are safe to assume that the advice and information in this book are believed to be true and accurate at the date of publication. Neither the publisher nor the authors or the editors give a warranty, expressed or implied, with respect to the material contained herein or for any errors or omissions that may have been made. The publisher remains neutral with regard to jurisdictional claims in published maps and institutional affiliations.

This Palgrave Macmillan imprint is published by the registered company Springer Nature Switzerland AG.
The registered company address is: Gewerbestrasse 11, 6330 Cham, Switzerland

If disposing of this product, please recycle the paper.

To Monica, Lorenzo, and Amelia—you are a part of this.

Acknowledgments

The period between my PhD defense and the submission of this book's manuscript has been a unique opportunity to reflect on the more than three years I have dedicated to this research and the exposure I have had to numerous voices striving to bring meaningful change to their contexts. This book attempts to amplify their work, and the narrative style aims to honor their contributions while producing research relevant for others. I am grateful to all the voices included in this work.

I also thank my PhD supervisors, Antonio Longo (Politecnico di Milano) and Joaquín Sabaté (Universitat Politècnica de Catalunya), for their invaluable guidance, as well as the series' editors, reviewers, commentators, and the PhD program's board and coordinators, who provided critical feedback and support throughout the process.

My doctoral research was co-funded by the PON REACT-EU program through the Italian Ministry of University and Research and included visiting stays at RWTH Aachen University and Universitat Politècnica de Catalunya. I thank Frank Lohrberg, Joaquín Sabaté, and their teams for their support during these research visits.

Finally, I thank my colleagues for their encouragement and, most of all, my family for their unwavering support throughout my academic journey.

Competing Interests The author declares the following as potential competing interests: The author completed a research fellowship at the Fondazione Patrimonio Ca' Granda as part of the PhD curriculum. This fellowship was a pre-arranged, mandatory activity. The foundation is mentioned in the Milan case study. The author did not receive any financial support from this institution.

Contents

About the Author

Antonio José Salvador is a Postdoctoral Research Fellow at the Department of Architecture and Urban Studies (DAStU) at Politecnico di Milano. He holds a PhD in Urban Planning, Design, and Policy, focusing on *Agrosocial Urbanism* and the governance of metropolitan agrarian landscapes in Milan, Barcelona, and the Ruhr Metropolis. He obtained a Master of Arts in Architecture from the Royal Danish Academy of Fine Arts and a Bachelor of Architecture from the University of Notre Dame in the United States, with a minor in Latin American Studies. From 2016 to 2019, he worked at the Metropolitan Institute of Urban Planning of Quito (IMPU). Earlier, he practiced as an architect and planner in the United States, Ecuador, and the Middle East. He has taught architecture and urban planning in both Italy and Ecuador. His publication record includes books, book chapters, and peer-reviewed articles on topics related to peri-urban territories, agrarian urbanism, metropolitan governance, large-scale projects, spatial planning, and related instruments and processes.

List of Figures

List of Tables

CHAPTER 1

Introduction

This publication brings together the outcomes of recently concluded doctoral research that examines how metropolitan regions are rethinking agriculture, land, and food systems in the context of the socio-ecological transition of territories and communities. Focusing on three European metropolitan contexts—Milan, Barcelona, and the Ruhr Metropolis—the study investigates how planning, landscape management, and governance practices have engaged with agriculture from the mid-twentieth century to the present, not merely as a residual type of land use to be protected but as a strategic component of contemporary metropolitan territories.

At the core of the book lies an exploration of peri-urban agriculture and its evolving role within local food systems. The research moves beyond a narrow understanding of farmland preservation—characteristic of earlier protection-oriented strategies—to examine how agricultural spaces are shaped by, and in turn shape, governance frameworks, land tenure arrangements, farm business models, ecological transitions, and social relations. Across the three case studies, agriculture emerges as a multifunctional practice capable of engaging with biodiversity protection, water and soil management, cultural heritage, social inclusion, and local economies. From this perspective, planning is understood not merely as a regulatory apparatus but as a mediating practice capable of articulating relationships between urban and rural areas, institutions and farmers, and production and consumption.

© The Author(s), under exclusive license to Springer Nature Switzerland AG 2026

A. J. Salvador, *Metropolitan Food Systems*, Sustainable Urban Futures, https://doi.org/10.1007/978-3-032-26016-1_1

The case-study structure of the book does not imply a direct comparative analysis. Instead, it foregrounds the specificity of each metropolitan region, showing how distinct historical trajectories, institutional settings, and territorial configurations generate different challenges and opportunities. At the same time, the research reveals shared trajectories, as these contexts have often been informally engaged in processes of knowledge exchange. Barcelona illustrates the tensions between long-standing horticultural systems, land insecurity, and collective initiatives that have laid the foundations for planning and management mechanisms that continue to serve as references for the protection of peri-urban agrarian territories and the articulation of local food networks. The Ruhr Metropolis highlights processes of adaptation and place resignification within a highly urbanized, post-industrial landscape, where multifunctionality, diversification, and design-led strategies play a central role, and where food production is increasingly prioritized by local and regional institutions as well as by a new generation of farmers. Milan exemplifies the growing influence of public institutions and large landowners in shaping land management, tenancy models, and food policy agendas, particularly in strengthening urban-rural linkages. At the same time, this case reveals a constellation of initiatives operating across different levels of government and planning processes that continue to struggle to establish more direct and coordinated relationships.

Methodologically, the book is grounded by a qualitative approach through extensive fieldwork, including site visits, interviews with farmers, planners, public officials, and local actors, as well as photography and cartography, and the analysis of policy documents and statistical data. These materials are assembled into a series of thematic chapters addressing key dimensions of metropolitan food systems, such as access to land, resource management, generational renewal, institutional coordination, and the role of farms as social and ecological infrastructures. Rather than proposing a single model, the book presents a constellation of practices, initiatives, and governance agreements that together form a multiscalar assemblage of references.

The book is structured in three parts. Chapters 1 and 2 establish the theoretical framework and provide a literature review that intersects multiple disciplines to articulate a contemporary agenda addressing the transition toward local and sustainable food systems. This agenda is framed as a response to unsustainable models of production and consumption driven by profit-oriented growth and intensive industrial methods. Chapter 3 introduces the three case studies through a historical perspective on planning processes and mechanisms, while also highlighting the need to consider additional tools related to landscape management and policymaking. Chapter 4 defines the three cases through quantitative and qualitative data, offering an overview of agricultural production and farming activities and serving as a reference for the subsequent thematic Chaps. 5, 6, 7 and 8. These chapters address critical dimensions that must be incorporated into planning, management, and policymaking efforts seeking to mediate the transition toward more local, sustainable, and equitable food systems, and are presented through vignettes and trajectories. The final chapter brings the cases together by identifying key obstacles that hinder this transition, while also highlighting pathways for action.

Ultimately, the book argues for an integrated and place-based approach—one that acknowledges the diversity of actors involved, embraces institutional complexity, and places agriculture at the center of broader debates on climate change, territorial governance, and the future of metropolitan regions.

A Planetary Approach to the Metropolitan Countryside

The binary framing of urban-rural relations has been extensively discussed and contested in scholarly debates, particularly through approaches that highlight practices, phenomena, and dynamics that do not fit neatly within these categories, including more-than-human understandings of territories (Bathla, 2024). A planetary approach to urbanization and the concept of extended urbanization, introduced by critical urban scholars in the early twenty-first century (e.g., Brenner, 2013; Schmid, 2019; Schmid & Topalovic, 2023), have themselves been challenged by theories and empirical cases that question the unidirectionality of these processes and advance counternarratives of the *ruralization* of the urban (Gillen et al., 2022; Krause, 2013). Wang et al. (2023) argue that many of these responses to the theory of planetary urbanization focus primarily on the "urbanization" component of Brenner's framework, while overlooking its "planetary" dimension. They therefore call for a planetary understanding of *ruralization* and a more-than-human perspective on extended urbanization processes that go beyond human consumption patterns. From this standpoint, urbanization should be understood as unfolding "around the globe, across city and countryside, and above and below the Earth's surface," through soil, particles, viruses, plants, and animals (Wang et al., 2023, p. 396).

This line of argument is further developed through a deconstruction of the urban-rural relationship as a product of Westernized academic traditions in urban and rural studies. Within an anglophone scholarly environment, the term *rural* often fails to capture social and cultural realities in other geographical contexts (Gkartzios et al., 2020). For instance, in Catalunya and other Catalan-speaking territories, the term *pagès* refers to the custodian of the countryside, of land cultivation, and food production. Its translation as *farmer* only partially conveys the cultural, social, and symbolic meanings embedded in the original term. Gkartzios et al. (2020) argue that *rural* has evolved into an English-oriented construct that has contributed to reinforcing the urban-rural binary. Despite sustained scholarly efforts to demonstrate the permeability of this divide, it continues to be socially reproduced, including through media representations of contemporary crises—particularly environmental ones—where urban and rural struggles are portrayed as relational yet framed in differentiated and often oppositional ways. Several authors (e.g. Gkartzios et al., 2020; Schmid & Topalovic, 2023; Woods, 2022) have therefore called for a planetary understanding of climate change and sustainability that extends beyond city-centric perspectives.

Another strand of critique highlights the persistent lack of recognition of *rural agency*, particularly in contexts in the Global East and South, where the urban-rural binary was imposed through colonial processes and Western epistemologies. In many of these territories, such distinctions did not previously exist and were introduced as part of colonial governance structures. This imposition reinforced settler-colonial relationships and racialized urban-rural hierarchies, in which cities were established for white settlers, while Indigenous and often nomadic communities were relegated to the rural, framed as underdeveloped or backward (Wang et al., 2023). In this sense, the urban-rural binary not only struggles to adequately represent contemporary Western societies, but also proves deeply inadequate for understanding other geographical and cultural contexts (Gkartzios et al., 2020).

In response, Wang et al. (2023) propose the notion of plural "planetary rural geographies" as both a conceptual vocabulary and a research approach. Within this framework, rural places are understood as assemblages of human and nonhuman actors, interconnected across urban and rural spaces and across local and global scales. This relational understanding of *ruralness* resonates with what Woods (2022) defines as the "global countryside," a term that informs the concept of the *metropolitan countryside* (da Silva Machado & Halfacree, 2022). Here, local narratives are interpreted through their entanglement with planetary processes and struggles. This approach seeks to reconcile the metropolis and the countryside—often perceived as incongruent geographical forms—by moving beyond interpretations of peri-urban areas as merely liminal zones, fringes, or transitional spaces (Di Campli & Gabbianelli, 2022). Instead, these hybrid territories are understood as sites where differentiated rural narratives persist and where the future of a countryside embedded within metropolitan regions can be actively contested and rearticulated (Gkartzios et al., 2022).

Agrarian Urbanism and the Agro-Urban Form

The concept of *agrarian urbanism*, retained and further developed by this book, re-emerges within the dialectics of contemporary planetary struggles. It is situated between the frameworks of "planetary urbanization" and "planetary rural geographies" (Brenner, 2013; Ho & Zhang, 2022; Schmid & Topalovic, 2023; Wang et al., 2023), and is mobilized to revisit the agrarian question within the complexity of global networks (Cuppini & Peano, 2019; Ghosh & Meer, 2021), while grounding the analysis in geographically situated and context-specific cases (Marshall, 2024; Topalović et al., 2023).

This perspective seeks to bridge urban and rural studies with planning theory and practice, and to address both the transformation of productive methods and shifts in consumption patterns grounded in solidarity, locality, and sustainable farming practices (Federici & Haiven, 2009; Tornaghi & Dehaene, 2020). While approaches to agrarian urbanism may emerge from diverse disciplines, geographies, and historical trajectories, they share a common emphasis on the *co-constitutiveness* of the agrarian and the urban as a fundamental condition for transcending the urban-rural divide.

This relational condition is not always intentional or formally articulated. Rather, it often unfolds through everyday actions and practices and may manifest through entangled and conflictual relationships or contested spatial forms (Cloke & Little, 1997; Woods, 2022). In this sense, agrarian urbanism is understood not as a fixed model, but as an evolving and negotiated condition through which agro-urban forms are continuously produced and redefined (Bathla & Lazzarini, 2025).

The Peri-Urban Question

This section examines the specificity of the peri-urban metropolitan countryside as an understudied geography—particularly from the perspective outlined in this chapter's introduction and within the growing agenda to articulate agrarian urbanism in the context of extended urbanization. Terry McGee's (1991)[1] work on Southeast Asian cities remains one of the most influential long-term analyses of peri-urban phenomena, conceptualizing the urban and the rural as co-constituted through economic, social, spatial, and environmental dimensions, as well as through the practices and livelihoods of those inhabiting these territories. McGee coined the term *desakota* and *desakotasasi*[2] to describe relational metropolitan landscapes in which agricultural production, industry, rapid urbanization, and formal and informal economies intersect. Since its introduction, the concept has evolved and has been employed to avoid overly generalized interpretations of the peri-urban, particularly in the study of Southeast Asian urbanization processes (McGee, 2002).

In the European context, a substantial body of urban studies emerged in the late twentieth and early twenty-first centuries, emphasizing the need for new analytical scales to understand cities in response to their spatial, social, economic, and environmental expansion (Font Arellano, 2007; Indovina, 1990; Viganò et al., 2018). Parallel strands of research have foregrounded the hybrid and contested nature of peri-urban areas, often highlighting their marginal or neglected status (Gallent, 2006) and focusing on the pressures generated by urban expansion, particularly the encroachment upon agricultural and natural land (Donadieu, 2013). Unsurprisingly, many planning and policymaking approaches inspired by twentieth-century paradigms have primarily sought to mitigate these pressures through containment strategies and farmland protection (Zasada,

[1]It refers to Terry McGee's 1991 book 'The emergence of desakota regions in Asia: expanding a hypothesis.'

[2]A concept created with Bhasha Indonesian words to describe a physical space, the in-betweenness of peri-urban regions where the coexistence of agricultural and non-agricultural activities occurs in extended zones. It was initially employed to study Indonesian urban regions and later employed in other Asian countries such as China, Vietnam, and the Philippines. When referred to it as a phenomenon, McGee employs the term *desakotakasi*.

2011). These approaches, however, have tended to devote less attention to placemaking processes (Gallent, 2006) or to recognizing the productive dimensions and agrarian structures of these territories (Ferraresi & Rossi, 1993; Sabaté Bel, 2015).

As a result, the peri-urban has often been framed as a third space—an in-between or hybrid territory functioning mainly as a zone of exchange. This perspective risks assigning borders and thickness to fringe areas (Gaeta, 2018) while obscuring the socio-spatial dynamics that actively shape these territories (Salvador, 2024; Schwarz & Streule, 2016). In contrast, relational approaches—such as those employed by McGee (2002) in Southeast Asia or Marshall (2024) in Mumbai—engage more deeply with the multidimensional complexity of peri-urban territories. Within these frameworks, agriculture emerges not merely as residual land use but as practice, livelihood, economic activity, and integral component of landscape structure and food provision systems. Raising the peri-urban question within an agenda for agrarian urbanism thus requires moving beyond liminal interpretations toward a more relational understanding of these geographies.

Such approaches resonate with metabolic and relational frameworks that planning practice has intermittently sought to conceptualize, including agglomerations and city-regions that integrate food production (Burke, 1966), bioregional models (Atkinson, 1992; Fanfani & Matarán Ruiz, 2020), and historically grounded concepts such as *foodsheds* (Kloppenburg et al., 1996). This chapter synthesizes the foundations for defining metropolitan agrarian territories within the broader context of planetary urbanization, the ruralization of food production and consumption, and the growing interest in local food systems. It situates this discussion within planning processes—further examined in Chap. 3—that have long sought to preserve farmland, while arguing that a planetary perspective, including more-than-human approaches, reveals a more complex and nuanced landscape for socio-ecological transition.

As argued throughout this chapter, planning can play a crucial role in this transition, not only by integrating insights from scholarly research but also by adopting appropriate scales of inquiry and engaging with integrative planning models. The following chapters will argue how these models should encompass land management, policymaking, governance structures, and a holistic understanding of resources and communities—dimensions that will be illustrated through the selected case studies.

References

Atkinson, A. (1992). The urban bioregion as "sustainable development" paradigm. *Third World Planning Review, 14*(4), 327–355. https://doi.org/10.3828/twpr.14.4.q1086507jk68vjvh

Bathla, N. (2024). Transhumance urbanism as an urban otherwise: Inhabiting agrarian incompletion at the intersections of extended urbanisation-extended Ruralisation. *Antipode, 56*(6), 2000–2020. https://doi.org/10.1111/anti.13085

Bathla, N., & Lazzarini, L. (2025). Towards an agrarian urbanism beyond the rural–urban divide. *Landscape Research, 50*(8), 1261–1271. https://doi.org/10.1080/01426397.2025.2591265

Brenner, N. (Ed.). (2013). *Implosions/explosions. Towards a study of planetary urbanization.* JOVIS. https://doi.org/10.1515/9783868598933

Burke, G. L. (1966). *Greenheart Metropolis: Planning the Western Netherlands.* Palgrave Macmillan. https://doi.org/10.1007/978-1-349-81771-9

Cloke, P., & Little, J. (1997). *Contested countryside cultures: Rurality and socio-cultural marginalisation.* Routledge.

Cuppini, N., & Peano, I. (Eds.). (2019). *Un mondo logistico: Sguardi critici su lavoro, migrazioni, politica e globalizzazione.* Ledizioni. https://doi.org/10.4000/books.ledizioni.11180

da Silva Machado, F., & Halfacree, K. (Chairs). (2022). For emancipatory rural futures in the metropolitan countryside(s) [Conference session]. RGS-IBG Annual International Conference 2022, Newcastle upon Tyne, United Kingdom.

Di Campli, A., & Gabbianelli, A. (Eds.). (2022). *Delinking. Lo spazio della coesistenza.* Lettera Ventidue.

Donadieu, P. (2013). *Campagne urbane. Una nuova proposta di paesaggio della città* (M. Mininni, Ed.; Vol. 102). Donzelli Editore.

Fanfani, D., & Matarán Ruiz, A. (Eds.). (2020). *Bioregional planning and design: Volume I: Perspectives on a transitional century.* Springer International Publishing. https://doi.org/10.1007/978-3-030-45870-6

Federici, S. (2009). On capitalism, colonialism, women and food politics. *Politics and Culture,* (2). Special issue on Food (&) Sovereignty https://www.campusincamps.ps/wp-content/uploads/2012/08/food_sovereignty-specissue.pdf

Ferraresi, G., & Rossi, A. (Eds.). (1993). *Il parco come cura e coltura del territorio. Un percorso di ricerca sull'ipotesi del parco agricolo.* Grafo edizioni.

Font Arellano, A. (Ed.). (2007). *La explosión de la ciudad: Transformaciones territoriales en las regiones urbanas de la Europa Meridional.* Ministerio de Vivienda.

Gaeta, L. (2018). *La civiltà dei confini. Pratiche quotidiane e forme di cittadinanza.* Carocci Editore.

Gallent, N. (2006). The rural–urban fringe: A new priority for planning policy? *Planning Practice & Research, 21*(3), 383–393. https://doi.org/10.1080/02697450601090872

Ghosh, S., & Meer, A. (2021). Extended urbanisation and the agrarian question: Convergences, divergences and openings. *Urban Studies, 58*(6), 1097–1119. https://doi.org/10.1177/0042098020943758

Gillen, J., Bunnell, T., & Rigg, J. (2022). Geographies of ruralization. *Dialogues in Human Geography, 12*(2), 186–203. https://doi.org/10.1177/20438206221075818

Gkartzios, M., Gallent, N., & Scott, M. (2022). *Rural places and planning. Stories from the global countryside*. Bristol University Press.

Gkartzios, M., Toishi, N., & Woods, M. (2020). The language of rural: Reflections towards an inclusive rural social science. *Journal of Rural Studies, 78*, 325–332. https://doi.org/10.1016/j.jrurstud.2020.06.040

He, S., & Zhang, Y. (2022). Reconceptualising the rural through planetary thinking: A field experiment of sustainable approaches to rural revitalisation in China. *Journal of Rural Studies, 96*, 42–52. https://doi.org/10.1016/j.jrurstud.2022.10.008

Indovina, F. (1990). *La citta diffusa*. IUAV. Dipartimento di analisi economica e sociale del territorio.

Kloppenburg, J., Hendrickson, J., & Stevenson, G. W. (1996). Coming in to the foodshed. *Agriculture and Human Values, 13*(3), 33–42. https://doi.org/10.1007/BF01538225

Krause, M. (2013). The Ruralization of the world. *Public Culture, 25*(2 70), 233–248. https://doi.org/10.1215/08992363-2020575

Marshall, V. J. (2024). *Periurban cartographies. Kolkata's ecologies and settled Ruralities*. ORO Editions.

McGee, T. G. (2002). Reconstructing "The Southeast Asian City" in an era of volatile globalization. *Asian Journal of Social Science, 30*(1), 8–27.

Sabaté Bel, J. (2015). Reflexiones en torno al proyecto urbanístico de un Parque Agrario. In *El Parque Agrario: Una figura de transición hacia nuevos modelos de gobernanza territorial y alimentaria* (pp. 93–112). Heliconia. http://hdl.handle.net/2117/83030

Salvador, A. J. (2024). Unpacking Parco Agricolo Sud through a socio-spatial production perspective. *TERRITORIO*, (2023/104). https://doi.org/10.3280/TR2023-104007.

Schmid, C. (2019). Analysing extended urbanisation. In *Future cities laboratory. Indicia 02* (pp. 157–180). Lars Müller Publishers.

Schmid, C., & Topalovic, M. (Eds.). (2023). *Extended urbanisation*. Tracing Planetary Struggles.

Schwarz, A., & Streule, M. (2016). A transposition of territory: Decolonized perspectives in current urban research. *International Journal of Urban and*

Regional Research, 40(5), 1000–1016. https://doi.org/10.1111/1468-2427.12439

Topalović, M., Hahn, D., Walter, J., & Westerheide, J. (2023). Agrarian project: Commoning land and labour around Zurich. *Studio Series New Ecologies.* https://doi.org/10.3929/ethz-b-000602332

Tornaghi, C., & Dehaene, M. (2020). The prefigurative power of urban political agroecology: Rethinking the urbanisms of agroecological transitions for food system transformation. *Agroecology and Sustainable Food Systems, 44*(5), 594–610. https://doi.org/10.1080/21683565.2019.1680593

Viganò, P., Cavalieri, C., & Barcelloni Corte, M. (Eds.). (2018). The Horizontal Metropolis: A Radical Project. In The horizontal Metropolis between urbanism and urbanization (pp. 1–9). Springer International Publishing. https://doi.org/10.1007/978-3-319-75975-3_1.

Wang, C.-M., Maye, D., & Woods, M. (2023). Planetary rural geographies. *Dialogues in Human Geography*, 20438206231191731. https://doi.org/10.1177/20438206231191731

Woods, M. (2022). Assets and assemblage in the global countryside. *Dialogues in Human Geography*, 20438206221075702. https://doi.org/10.1177/20438206221075702

Zasada, I. (2011). Multifunctional peri-urban agriculture—A review of societal demands and the provision of goods and services by farming. *Land Use Policy, 28*(4), 639–648. https://doi.org/10.1016/j.landusepol.2011.01.008

CHAPTER 2

Defining Local Food Systems

Introduction

The intention behind this chapter is twofold. First, it introduces an interdisciplinary body of literature and academic debate on local food movements and local food systems from various social sciences and primarily from fields of study that place agriculture, food, and rural places at the core of their research. Second, it narrates the evolution of agricultural production in dialogue with social, environmental, and economic concerns from the 1990s until present times. This is a moment in which family-led farms and communities deliberately seek to challenge the "corporate food regimes," a hyper-globalized consumption model associated with the so-called Green Revolution (Goodman & Watts, 1997; Holt Giménez & Shattuck, 2011). A model with significant social and environmental implications and a cultural impact on how communities relate to food production and consumption, transforming diets and lifestyles across the globe (Altieri & Toledo, 2011; Whatmore & Thorne, 2004).

The chapter follows a linear structure to emphasize the evolution of various discourses such as agricultural production, business models, and food systems, while reinforcing academia's role in tracking these changes and linking them to a broader global discourse on sustainability and climate change, and multiple crises. It also intentionally links the academic debate led by the social sciences in bringing forward the *food question*—as a key pillar of socio-ecological transition of communities and places—with

© The Author(s), under exclusive license to Springer Nature Switzerland AG 2026
A. J. Salvador, *Metropolitan Food Systems*, Sustainable Urban Futures, https://doi.org/10.1007/978-3-032-26016-1_2

planning, design, and governance mechanisms discussed in later chapters. The analysis also shows how, in the 2000s, landscape approaches to food and agriculture (e.g., Vejre et al., 2007) made valuable contributions that have since lost ground to policymaking framed by urban governance. This chapter takes an essential step toward bringing the planning and design disciplines into closer engagement with these underexplored themes in urbanism and urban studies, elevating them as priorities for both professional practice and urban governance.

Rural Development and Multifunctionality in Agriculture

Until the 1990s, agricultural development across Europe was primarily guided by modernization principles that emphasized scale enlargement, intensification, specialization, and industrialization (van Der Ploeg & Roep, 2003). Supported by regional and national policies within an increasingly globalized context, this agri-industrial model expanded rapidly across Europe and beyond. However, this model led to significant socio-economic and environmental consequences, including the decline in farm numbers, loss of employment opportunities, and a resulting rural exodus (Marsden, 1999). It also intensified tensions between agricultural production and environmental sustainability (Knickel, 1999), while widening regional disparities and threatening traditional rural lifestyles and landscapes (Marsden, 1999). These transformations eventually opened the possibility to question the agri-industrial model itself and its long-term implications for rural communities.

In *Rural Futures: The Consumption Countryside and its Regulation* (1999), Terry Marsden attempts to draft a new research agenda for rural territories, urging to overcome critiquing the agri-industrial model toward exploring diverse and multifunctional alternatives in food systems and agricultural practices. Marsden emphasized the rise of bottom-up initiatives and critiques uniform, state-centrist rural policies across Europe. Marsden called for locally grounded, spatial approaches to capture the complexity of rural spaces within what he defined as the "contemporary consumption countryside." This emerging paradigm, he argues, redefined rural development as a socially and politically constructed process shaped by new economic relations of production, exchange, and trust, with differentiated effects at both societal and local scales, specifically within rural localities (Marsden, 1999, p. 503).

Among the factors reshaping rural societies were the decline of agricultural employment, structural transformations in farming, the rise of ethical and political environmentalism, demographic shifts, new societal expectations toward land and landscape, reduced public support, and expanding free trade. Therefore, Marsden's notion of the *consumption countryside* depicts an uneven rural landscape shaped by external relationships and a wide range of demands for adapting and innovating. Within this context, farms—as the basic unit—and local communities emerged as key actors in guiding integrated rural development and informing policymaking.

The Emergence of Rural Development as a New Paradigm

As changes unfolded during the 1990s, various publications from different fields began identifying *rural development* as a societal model and *multifunctionality* as an economic model responding to the failures of agricultural modernization. Van der Ploeg and Roep (2003) argued that these became key concepts in steering European agriculture away from the modernist paradigm. Economic pressures—such as the milk quota system, declining commodity prices, and rising energy costs—along with concerns about environmental sustainability, food safety, and animal welfare, the integration of Central and Eastern European countries into the EU, and reduced public support, played a crucial role in fostering rural development initiatives (van Der Ploeg & Roep, 2003; Ward, 1993). Family farms became central advocates of this shift, seeking alternatives to industrialization and large-scale farming. Van der Ploeg and Roep (2003) further argued that the transition from agricultural modernization toward rural development should have served as the guiding principle for policy design, enterprise development, and institutional arrangements. This is particularly relevant because, while multifunctionality from a socio-economic perspective entails a reconstruction of the economic base of farms directly as enterprises, when understood as part of a broader paradigm shift—namely, rural development—it entails new relationships with governance and management entities across scales, yet remains grounded in historical traditions.

The Role of Multifunctionality in Rural Development

Multifunctionality is a framework that integrates diverse activities into agriculture, including sustainable farming, landscape management, cultural and touristic initiatives, and energy production (van Der Ploeg & Roep, 2003). Studies emphasize the importance of creating "synergy" between agricultural production and other rural economic activities (Brunori & Rossi, 2000; Knickel & Renting, 2000). In this context, synergy refers to mutually reinforcing relationships among different activities that strengthen rural economies and enhance their sustainability. This concept marks a key distinction between rural development and modernization. While modernization promotes specialization and the segregation of agriculture from other rural activities, rural development seeks to integrate and generate reciprocal benefits among diverse agricultural and non-agricultural activities (van der Ploeg & Roep, 2003).

These arguments are significant because they recognize the complexity of farming across scales and levels, the structural diversity of agricultural systems, and their potential to serve broader social purposes beyond the farm economy. Texts such as Brunori and Rossi (2000) and de Roest (2000) highlight synergies between scales—or levels, as commonly referred to in the social sciences—linking local artisanal production with regional experiences in a globalized and industrialized context. Guzmán et al. (1999) also attempt to connect agroecology as a model within the broader framework of sustainable rural development. These two themes would later re-emerge more consistently in the following decades.

One of the main contributions of van der Ploeg and Roep (2003) was the identification of five key levels of rural development, which are also central to understanding the complexity and spatial dimensions of food systems:

- Global interactions between agriculture and society
- A new developmental model for the agricultural sector
- Individual farm households and their economic sustainability
- The countryside and its economic actors
- Policies and institutions shaping agricultural practices

The Transformation of Farms Into Rural Enterprises

The notion of farmers as *rural entrepreneurs* was introduced by Marsden (1999), framing agriculture and the farming business model as needing to adapt to the changing demands of European society (van Depoele, 1996) and to overcome the economic pressures affecting European agriculture (van der Ploeg & Roep, 2003). This transition, already underway in the 1990s, led van der Ploeg and Roep (2003) to further conceptualize the evolving role of the farm.

As argued by van der Ploeg and Roep (2003), the farm as an enterprise encompasses three dimensions: (1) the agricultural dimension, representing traditional production; (2) the rural dimension, which includes landscape and cultural contributions; and (3) the mobilization and use of resources, involving land, labor, and capital. With the evolution of conventional agriculture, the farm evolves into a more complex enterprise that comprises new functions and activities, while rural development reconfigures the relationship among these dimensions through forms of cooperative management. Van der Ploeg and Roep (2003) developed a model to describe these relationships and the outcomes through three concepts: ***broadening, deepening, and regrounding***. This model illustrates how farms diversify their activities, improve product quality, and optimize resource use (van der Ploeg & Roep, 2003). Such transformations reflect a broader shift from production-oriented farming toward more complex rural enterprises engaged in multiple markets (Goodman & Watts, 1997; Robinson, 2003).

Although the individual farm emerged as a key unit of analysis, other studies focused on farm networks, collective practices, and shared characteristics across farms. One of the most influential publications in this regard is Whatmore's *Nourishing networks. Alternative geographies of food* (1997, 2004), first published in Goodman & Watts (eds.), *Globalising food. Agrarian questions and global restructuring*. This work seeks to map networks and new geographies employing the term ***alternatives***—a term that would become widely used in the following decade to describe an alter-ego model to corporate food regimes or industrial agriculture (Holt Giménez & Shattuck, 2011). This body of work would later refer to these collective practices as ***alternative food networks*** (AFNs) (Goodman et al., 2013; Sonnino & Marsden, 2006).

By the end of the twentieth century, rural development was presented as a multifaceted, multi-actor, and multilevel model in which synergies across scales and levels were key to fostering new forms of social cohesion. This emerging model once again recognized the role of farms—both as units and as enterprises—as central to this transformation, in contrast to the modernization paradigm that had regarded individual farms as superfluous (van der Ploeg & Roep, 2003).

During this decade, rural development and multifunctionality were increasingly understood as representing a paradigm shift in European agriculture, challenging market- and technology-driven modernization approaches. This shift comprised localized initiatives, new forms of urban-rural interactions, and diversified economic models. While national policies often struggled to integrate these principles, academic discourse progressively incorporated rural development principles through the observation of farming practices. The role of the farm therefore emerged as a central element, particularly in its interaction with the other levels defined in van der Ploeg and Roep's (2003) five-level model of rural development.

At the same time, this shift also introduced new challenges, including land-use conflicts, governance issues, and concerns regarding economic sustainability (Knickel & Renting, 2000). Nevertheless, as emphasized in academic debate, it also opened opportunities to realign agriculture with societal needs, strengthen rural–urban linkages, and ensure the long-term viability of rural economies.

Multifunctional Agriculture and Multifunctional Landscapes

With the evolving role of the farm and its multifunctional practices, a natural question emerged regarding its relationship to the broader agrarian landscape. To address this, several studies sought to reflect on the multiple meanings and applications of the term *multifunctionality*. Publications such as Brandt & Vejre (2004), Renting et al. (2009), and van Huylenbroeck et al. (2007) defined multifunctionality from different perspectives—some focusing on agricultural activities and land use as material expressions of societal demands, while others discussed the concept in relation to agricultural and environmental policies (Abler, 2004). Multifunctionality was also applied to describe landscapes themselves, emphasizing the relationship between the production of primary material goods (land-dependent activities such as agriculture or forestry) and other material and non-material benefits, including CO_2 sequestration and groundwater recharge (Zasada, 2011). This latter interpretation highlights a more metabolic relationship between ecosystems and human practices.

A key text from this period is Vejre et al. (2007) *Multifunctional agriculture and multifunctional landscapes - land use as an interface*. The authors sought to analyze and define the concept of multifunctionality within academic and professional debates, addressing the conceptual discrepancies that had emerged during the previous decade. Vejre et al. (2007) distinguished between two primary perspectives:

- Agricultural multifunctionality, which views farming as an economic activity producing multiple outputs beyond food, including recreation, aesthetics, and wildlife habitats.
- Landscape multifunctionality, which considers landscapes as inherently multifunctional systems that integrate economic, environmental, and social functions.

The publication by Vejre et al. (2007) adopts Forman and Godron's (1986) definition of *landscape ecology*, in which a landscape is understood as a physical spatial unit capable of fulfilling several purposes—or possessing multiple functions—for society. This approach is grounded in the idea that landscapes, by definition, consist of several spatial units that can each satisfy different purposes and functions.

Agriculture vs. Landscape: Different Analytical Approaches

In this period, there is also an attempt to establish a relationship between agriculture and landscapes, emphasizing how agricultural activities shape cultural landscapes and how landscapes, in turn, provide the context in which these activities take place (Vejre et al., 2007). However, as further argued by Vejre et al. (2007), the concept of multifunctionality introduces tensions between these approaches because the disciplines involved differ: Studies on agricultural multifunctionality are mainly rooted in agronomy, agricultural and environmental economics, and sociology, whereas research on landscape multifunctionality is primarily conducted within biology, geography, social sciences, and the humanities.

A principal argument presented by Vejre et al. (2007) is the fundamental difference in the analytical units used in agriculture and landscape studies. From the agricultural perspective, the farm serves as the primary economic and social unit for analyzing multifunctionality, whereas from the landscape perspective, multifunctionality is examined through the landscape as a complex land system, analyzed at different spatial scales. Vejre et al. (2007) further argued that the dialectics of multifunctionality extend beyond academic debate, as they are central to agricultural policy-making, international trade, and environmental protection, and that effective policy design requires a mutual understanding of these distinct approaches.

The "agricultural path" to multifunctionality defines it as an economic activity with multiple outputs. Agricultural products and production methods support other functions, including recreation, aesthetics, and habitat. Vejre et al. (2007) emphasize that multifunctionality is often perceived as a neutral characteristic of agriculture. However, when employed as a normative concept, it is used to value outputs beyond agricultural goods, such as landscapes, employment, and wildlife. In this perspective, multifunctionality is activity-oriented, where agriculture remains the primary function, while other product externalities are considered secondary functions.

In contrast, the "landscape path" to multifunctionality focuses on the ecological aspects of landscapes to understand their variations and the consequences of these variations. Vejre et al. (2007) argue that this approach acknowledges multifunctionality as an intrinsic property of landscapes. They further suggest that interest in landscape multifunctionality arose because, in the 1990s, many environmental problems in rural areas were linked to the effects of monofunctional, industrialized agricultural landscapes that eliminated non-productive land uses. This situation underscored the need for multifunctional strategies integrating diverse land uses. Forman and Godron (1986) define function as the interaction of spatial elements, highlighting the importance of integrating different land uses to maintain ecological and social balance. Within this approach, multifunctionality is considered an emergent feature in which all functions are treated equally, though the emphasis depends on the observer's focus. The landscape perspective primarily examines the relationships between landscape functions.

Land Use and Landscape Management as Binding Elements

For Vejre et al. (2007), land use serves as the element that binds the agricultural and landscape approaches to multifunctionality. Agricultural production often constitutes the primary function of open landscapes, and the farm represents agriculture as a system or network embedded in a social and spatial context. Its primary function is agricultural production, and the land devoted to this activity constitutes a defined land use, which simultaneously becomes a landscape function within a broader landscape system. In this way, land use functions as a common denominator, connecting the farm system with the surrounding rural landscape and situating an economic activity within a wider social and environmental context.

Among Vejre et al. (2007)'s main contributions lie not only the conceptualization of multifunctionality through these two approaches but also the demonstration of how farm and landscape systems interact through agricultural activity, and how this activity is transformed into a landscape function, relating to broader social and ecological processes. Furthermore, they highlight that a common operator—the management of agricultural landscapes—can integrate both approaches. By introducing a *management* component, Vejre et al. (2007) draw on Hägerstrand's (1995) concepts of territorial, spatial, and functional competencies, emphasizing the importance of integrating the competencies of individuals with the spatial competencies exercised by authorities through various governance and planning mechanisms. This perspective supports an operational approach to landscape multifunctionality in which knowledge is embedded in tangible tools both at the societal level (e.g., planning and policy design) and at the farm level (e.g., advisory, management, and planning tools). The questions raised by Vejre et al. (2007) further challenge the notion of multifunctionality as a neutral term and highlight its relationship with sustainability through the evaluation of its functions.

Reconsidering Multifunctionality

With multifunctionality becoming an established focus of study across various fields, emerging research at the turn of the century examined specific aspects, including lifestyle farming, in which peri-urban farms act as attractors of a "functionally urbanized countryside" (Præstholm & Kristensen, 2007). In these cases, despite the predominant agricultural land use, farms host other economic activities, with leisure often becoming more prominent than production (Præstholm & Kristensen, 2007). Other studies focused on multifunctionality with environmental and climate-related benefits, such as agriculture contributing to carbon sequestration, water management, flood control, and urban climate moderation (e.g., Freibauer et al., 2004; Haase & Nuissl, 2007; Lamptey et al., 2005).

Social and cultural aspects also gained attention, recognizing agriculture as an integral component of cultural landscapes capable of integrating aesthetic and environmental functions (e.g., Hall et al., 2004; Sabaté Bel & Benito del Pozo, 2010), especially as urban visitors increasingly perceive agricultural landscapes as recreational and aesthetic spaces rather than purely productive ones (Buijs et al., 2006).

At the same time, authors such as Belletti et al. (2003) began to question the relationship between multifunctionality and broader rural development, challenging the capacity of agriculture to address wider socio-economic issues beyond transforming rural spaces into sites of consumption (Gallent, 2006). This detachment, perhaps driven by local government and policymaking interests, often reinforced the economic dimension of agriculture through regulatory frameworks, emphasizing its role as an enterprise capable of producing both commodity and non-commodity outputs (Vandermeulen et al., 2006). One particularly evident strategy, mainly in peri-urban areas, involves farmers' engagement in landscape management activities, which also provides important economic support (Van Huylenbroeck et al., 2007). There has also been an emerging understanding of the transformation of the countryside within the context of metropolitan areas, explored through specific case studies, such as Busck et al. (2008); Jongeneel et al. (2008); Vandermeulen et al. (2006). Many of these studies focused on northern European cities and metropolitan regions, while others, such as Aguglia et al. (2009), examined multiple contexts, including Italy and the Netherlands, to understand how multifunctionality evolves in different national settings.

One of the key contributions of Aguglia et al. (2009) is their categorization of multifunctionality at various levels (pg.3): (1) Farm level, where agricultural resources are used for non-market outputs such as biodiversity conservation; (2) Rural business level, involving diversification into activities like energy production and tourism; and (3) Household level, reflecting pluriactivity and the use of family resources for multiple income streams. The same study further classifies multifunctionality into four types of functions, following Van Huylenbroeck et al. (2007): *green functions*, for landscape and biodiversity management; *blue functions*, for water resource management and flood control; *yellow functions*, for rural vitality, historical and cultural heritage, and amenities; and *white functions*, for food security and safety.

Aguglia et al. (2009) conclude that farm multifunctionality has been primarily driven by a combination of incentives: formal ones, such as regulations and policies; informal ones, guided by shared values; and market-based incentives, responding to consumer expectations. They further argue for overcoming the perception of multifunctionality merely as a survival strategy, advocating instead for seeing farms as businesses where land is a resource not solely for food production or commodities, but also for raw materials, fuel and energy, services, and other non-commodity outputs.

Policy Approaches for Revisiting Urban-Rural Linkages

The emerging role of local governments and policymaking has also gained significant attention in fostering new urban-rural relationships by establishing a more direct nexus between producers and consumers (Buciega et al., 2009). Although multifunctionality initiatives led by farmers included direct sales and the opening of their farms to other commercial functions and activities, municipalities played a key role in facilitating public urban spaces, such as farmers' markets, where local producers could gain direct access to consumers (Chiffoleau, 2009). Some farmers found a niche in specialized, high-value products (Zasada, 2011), which directly influenced their farming methods and type of production. However, authors such as Born & Purcell (2006) have warned of a potential *trap* in overpricing local products or adopting a consumer-oriented approach that may overshadow broader sustainability goals. Zasada's (2011) analysis characterized peri-urban farming as a "heterogeneous pattern of holdings with intensive and specialized production, high participation in diversification, and low-intensive hobby and lifestyle-oriented farms" (Zasada, 2011, p. 646).

The increasing number of alternative food networks (AFNs) has also led to the exploration of the complexity and diversity of these networks, primarily as emerging bottom-up initiatives driven by both consumers and producers, and to the recognition of their role in redefining food systems, direct producer-consumer relationships, and farming practices (Holloway et al., 2007; Whatmore & Thorne, 2004). Such networks, and the relationships they foster with farms and farmers, have been described as new "types of social contracts" (Holloway et al., 2007). For instance, these networks have strengthened models of community-supported agriculture (CSA), enabling farmers to adopt more sustainable farming methods while securing a committed group of consumers.

AFNs and other short supply chain-oriented activities have also reinforced the *household level*, as defined by Aguglia et al. (2009), whereby household members who were previously uninvolved in farming contribute to activities required by these new functions, including direct marketing (Jarosz, 2008). Authors such as Paül and McKenzie (2013) have sought to link AFNs with farmland conservation efforts, often guided by planning initiatives. These efforts are exemplified in Italy and Spain through governance and planning mechanisms, such as agricultural parks (to be explored in Chap. 3), which aim to integrate farmland protection tools with policies and strategies that promote urban-rural exchange through short supply chains and territorial marketing (Fanfani, 2006).

During this decade, the concept of multifunctionality evolved to encompass environmental, economic, and social dimensions, primarily guided by a *landscape approach* (Vejre et al., 2007). Vejre et al. (2007) introduced key analytical elements, particularly by distinguishing the units of analysis employed by different disciplines studying multifunctionality. For instance, within social sciences, the farm gained increasing relevance as it came to be understood as a rural enterprise with essential resources capable of producing various goods and services of significant public value.

This shift in perspective promoted greater contextual differentiation, particularly by recognizing peri-urban areas as key settings for observing the transformations occurring in agriculture and rural territories, as well as their evolving relationship with urban populations. Future research agendas highlighted the growing role of local policymaking and the challenge of integrating multifunctionality into governance mechanisms, ensuring that land-use planning simultaneously addresses agricultural, ecological, and social needs, and that the new rural development paradigm places agriculture as an attractive and competitive counterpart to urban development (Zasada, 2011).

Networks, Collectiveness, and Policy Integration in Local Food Systems

In the following decade, several studies sought to expand the tools available for agricultural land management, moving beyond generic containment strategies by explicitly considering the role of farms and agriculture (Perrin et al., 2020). A notable contribution is Kerselaers et al. (2011), *Prioritising Land to Be Preserved for Agriculture: A Context-Specific Value*

Tree, which establishes a dialogue between the farm scale and the broader farming area through a "value tree" framework. This framework assesses farmland value by combining farm-level qualities, such as sustainable practices, with land-level attributes, including soil quality, management, and agricultural vocation. The study argues that planning and policymaking should move beyond generic farmland preservation to prioritize land with higher productive potential and farms that implement more sustainable business models. A key dimension highlighted in this work—and one that gained increasing attention in the following decade—is the role of policymaking, which later evolved into a distinct field under the concept of *urban food policies* (Moragues-Faus & Morgan, 2015).

Other studies, such as Brinkley (2012), also emphasize the importance of generating evidence on the social, environmental, and economic benefits produced by individual farms as a key indicator for farmland preservation. Brinkley's work seeks to assign a market value to these benefits, highlighting that higher-value outcomes tend to occur where farmland preservation policies have promoted stronger synergies between agricultural land and nearby urban areas. More recently, Corsi et al. (2023) have explored the relationship between farm multifunctionality and the ecosystem services framework to evaluate agriculture's impacts beyond food and fiber production. Their analysis classifies these impacts into four categories: supporting services, such as biodiversity and soil formation; regulating services, including air quality, climate, and water regulation; provisioning services, such as food and raw materials; and cultural services, encompassing mental and physical health as well as aesthetic values.

Texts such as those by Dwyer (2011), Paül and McKenzie (2013), and Winter (2009) introduce perspectives that highlight the need to integrate additional dimensions into the understanding of farmland preservation, moving beyond land use. Dwyer (2011), for instance, emphasizes the importance of regional and local policymaking and the integration of climate, energy, food, and health policies within spatial planning. Winter (2009) underscores the role of data as a key resource for linking rural and environmental policies, particularly for assessing land-use change and multifunctionality in the context of climate change and the growing concern for food and energy security. Paül and McKenzie's (2013) study perhaps best exemplifies this integrative approach. Focusing on the Barcelona metropolitan area, the authors explore the relationship between farmland preservation, local food systems, emerging alternative food networks, and the specific challenges of peri-urban agriculture. Their work underlines farmland preservation as a fundamental condition for alternative food networks to individually and collectively thrive and interprets the dialectics between these dimensions as a form of rural-urban partnership in which planning and policymaking can mutually support one another.

A recent contribution by Perrin et al. (2020) reviews land policies aimed at farmland protection, with particular attention to peri-urban contexts. The authors note that policymaking often focuses on land rather than agriculture, highlighting the need for place-based approaches that reflect local dynamics and the specific nature of farming. The study acknowledges the extensive body of instruments for farmland preservation developed over the previous decade but stresses the importance of considering regional and local conditions to prevent land abandonment and support agriculture as an economic activity. Perrin et al. (2020) also underline the complexity of farmland protection as a policy issue, frequently marked by tensions with other social, political, and spatial agendas. Among the specific challenges of peri-urban agriculture identified are land fragmentation, soil erosion, and speculation. Issues that, as Zasada (2011) and Piorr et al. (2018) argue, call for agricultural policies tailored to the scale, diversity, and actors characteristic of peri-urban areas.

The Genealogy and Present Structure of Urban Agriculture

The development of an agenda for urban and peri-urban agriculture represents a major achievement, largely accomplished as the collective efforts of European scholars (Jansma et al., 2024; Lohrberg et al., 2015) and by the growing engagement with broader dimensions such as social movements, food justice, and heritage (Horst et al., 2017; Lohrberg et al., 2022; Tornaghi, 2014). Many of these contributions stem from EU-funded projects and the increasing interest in building a European framework for local food systems that acknowledges the role of local governance and urban agriculture. Projects such as COST Action Urban Agriculture Europe (Lohrberg et al., 2015) and Horizon 2020 initiatives including EFUA, Food Trails, and FUSILLI Urban Food Planning (Wascher et al., 2023) have become active research networks exploring urban agriculture across diverse contexts. The COST Action Urban Agriculture Europe project, in particular, sought to define the spatial and functional dimensions of urban agriculture—its motivations, actors, origins, production types, and consumer relationships—while developing typologies based on business models, scale, and production scope. It distinguishes *urban farming* from *urban food gardening*, the former encompassing activities such as educational, leisure, or social farms centered on on-site experiences, as well as environmentally driven or market-oriented production. Urban food gardening, by contrast, includes both individual initiatives, such as allotment and family gardens, and collective ones, such as community and educational gardens.

Building on this work, Jansma et al. (2024) expand the work by introducing new practices and defining six typologies: the urban farm, the community park, the do-it-yourself garden or farm, the zero-acreage farm, the social farm, and the community garden. These typologies also include emerging forms such as vertical and rooftop farming. Although primarily spatial in nature, describing the physical contexts in which agriculture takes place, they also account for the types of stakeholders involved, their scope, and the internal dynamics of each model. The methodological approach employed in both Lohrberg et al. (2015) and Jansma et al. (2024) reinforces the relevance of the individual farm holding within broader territorial and socio-economic contexts. In these more recent studies, however, the contribution lies in adopting a spatial perspective that situates agriculture within planning and design processes—and as a result of these processes.

The publication by Lohrberg et al. (2022) frames urban agriculture as a heritage practice, reconstructing through multiple contexts a genealogy of urban agriculture rooted in the historical evolution of cities and society. This perspective is essential to counter the perception of urban agriculture as a recent phenomenon. It also contributes to preserving the diverse scales, forms, and scopes of agricultural practices that take place within and around cities, emphasizing that agriculture and food production are integral components of urban life. The atlas developed by the authors addresses a significant scholarly gap by documenting urban agriculture as both a historical practice and a valuable resource for informing contemporary food policy and planning.

Networks, Assemblages, and an Agenda for the Socio-ecological Transition of Food Systems

A pivotal moment for food policymaking and governance occurred in 2015 with the Milan Expo, when the City of Milan launched the Milan Urban Food Policy Pact as part of its legacy. The pact positioned cities and local governments as central actors in developing sustainable food systems and has since grown into a global network, enhancing Milan's international visibility (Susi Botto & Di Vita, 2017). This development coincided with the UN Habitat III New Urban Agenda, which underscored the critical role of local governments in promoting sustainable urban living. Under this framework, several city networks emerged to exchange knowledge and support food-related policies. These *translocal networks*—as defined by Moragues-Faus and Sonnino (2019)—became arenas for practice-based innovation in urban food governance. Building on earlier work by Moragues-Faus and Morgan (2015), which first conceptualized urban food policies as complex governance systems involving local governments, civil society, and private actors, Moragues-Faus and Sonnino (2019) provided empirical evidence of how such networks operate. Their study of the Sustainable Food Cities Network in the UK revealed how these assemblages promote collective action, knowledge sharing, and new forms of political agency. Cities engage with translocal networks not only to co-produce shared imaginaries and challenge governance structures, but also to connect with broader social movements, highlighting the role of food and agriculture in mobilizing collective action.

For instance, peasant movements in Latin America, such as La Via Campesina have inspired urban food initiatives (Altieri & Toledo, 2011; Borras, 2023), illustrating how transnational exchanges shape local practices. This perspective also helps explain the evolution of agroecology from a scientific framework into a socio-political movement (Gentili & Zampetti, 2021; Wezel et al., 2009), offering insights into pathways for the socio-ecological transition of places, communities, and food systems (Akram-Lodhi, 2021; Tornaghi & Dehaene, 2020). Although the strategies to guide such transitions remain an open challenge, scholars have debated the core values underpinning them—solidarity, justice, and collectiveness (Haiven, 2009; Tornaghi & Dehaene, 2020)—and advocate for a convergence of diverse agendas, including political ecology, critical agrarian studies, agroecology, and social movements, progressively reframing the agrarian question into the food question and laying the groundwork for a political agroecology agenda linking agriculture, ecological crisis, and capitalism (Akram-Lodhi, 2021; Giunta, 2023).

The diversity of topics in this decade reflects the multiple strands and approaches to agriculture and food-related issues. The discourse shifts from a focus on agriculture and farmland to broader concerns about food and food systems, highlighting the growing role of cities and local governments in addressing food-related challenges. Scholarly research also expands across geographies, cultures, governance forms, disciplines, and historical periods, tracing the origins and evolution of practices across scales. This has revealed a highly complex universe of food production and consumption, which has become central to the socio-ecological transition of food systems, communities, and territories.

Conclusion

This chapter reveals the complexity of agrarian territories and farming as an economic activity that carries a strong cultural dimension, which has been progressively reconfigured by societal lifestyles and market-driven political decisions that have supported a production and consumption model with detrimental effects on the environment and on our cultural understanding of farming and food. This analysis also shows a clear attempt, mostly driven by family farmers, to resist and rearticulate production and consumption models that are not only the result of pursuing their economic sustainability, but are also capable of creating social value based on locality, diversity, and more sustainable farming methods.

The review also demonstrates that food production and agrarian territories play a crucial role in the socio-ecological transition of territories. Most recent policymaking and governance efforts attempt to reinforce these models, which emerged organically at the turn of the century. However, a key challenge for future research lies in understanding the conditions under which this transition should occur and the responsibilities shared among stakeholders. This raises important questions about the role of planning in supporting local food systems and managing agricultural territories. The analysis presented in this chapter informs the book's content by expanding the perspective of planning and design disciplines, particularly regarding governance, land management, and the sustainability of agricultural practices. Consequently, the following chapters build on the key topics identified here and presented in Table 2.1 to explore the processes and mechanisms that support the socio-ecological transition of agrarian territories and communities.

Table 2.1 Topics and sources emerging from Chap. 2

Main topic	*Subtopics*	*References*
Farming styles and business models	Multifunctionality	Aguglia et al., 2009; Marsden, 1999; Van der Ploeg and Roep, 2003; Brandt & Vejre, 2004
	Farming models	Guzmán et al., 1999; Van der Ploeg and Roep, 2003
	Individual farm households	Van der Ploeg and Roep, 2003; Aguglia et al., 2009
	Rural enterprises	Brunori & Rossi, 2000; de Roest, 2000; Knickel & Renting, 2000; Van der Ploeg and Roep, 2003; Whatmore & Thorne, 2004
	Urban agriculture	Jansma et al., 2024; Lohrberg et al., 2015
Production and supply chains	Alternative food networks	Goodman et al., 2013; Guzmán et al., 1999; Holloway et al., 2007; Sonnino & Marsden, 2006
	Food regimes	Holt Giménez & Shattuck, 2011
	Relocalization of food systems	Born & Purcell, 2006; Goodman et al., 2013
	Direct marketing	Chiffoleau, 2009; Holloway et al., 2007; Marsden, 1999
Agricultural land use	Landscape management	Hägerstrand, 1995; Vejre et al., 2007
	Multifunctional landscape system	Vejre et al., 2007
	Land use	Gallent, 2006; Vejre et al., 2007
	Peri-urban agriculture	Fanfani, 2006; Gallent, 2006; Piorr et al., 2018; Zasada, 2011
	Non-productive functions	Brandt & Vejre, 2004
	Farmland protection	Fanfani, 2006; Paül & McKenzie, 2013; Sabaté Bel & Benito del Pozo, 2010
Community cohesion	Farm networks	van Broekhuizen et al., 1997; Lowe et al., 1995
	Social contracts	Guzmán et al., 1999; Holloway et al., 2007; Jarosz, 2008
	Conflicts	Gallent, 2006; van der Ploeg and Roep, 2003; Zasada, 2011
	Social movements and food justice	Altieri & Toledo, 2011; Borras, 2023; Haiven, 2009; Holt Giménez & Shattuck, 2011; Tornaghi, 2014

(*continued*)

Table 2.1 (continued)

Main topic	*Subtopics*	*References*
Biodiversity and nonhuman agency	Agroecology	Gentili & Zampetti, 2021; Guzmán et al., 1999; Tornaghi & Dehaene, 2020; Wezel et al., 2009
	Sustainable farming	Holloway et al., 2007
	Biodiversity management	Van Huylenbroeck et al., 2007
	Socio-ecological transition	Akram-Lodhi, 2021; Giunta, 2023; Tornaghi & Dehaene, 2020
Resources management	Ecosystem services and functions	Corsi et al., 2023; Zasada, 2011
	Landscape ecology	Forman & Fodron, 1986; Vejre et al., 2007
	Metabolism	Freibauer et al., 2004; Haase & Nuissl, 2007; Lamptey et al., 2005
	Cultural landscapes	Hall et al., 2004; Sabaté Bel & Benito del Pozo, 2010
Land tenure regimes and management models	Land prices	Robinson, 2003
	Farming contracts	Perrin et al., 2020; Zasada, 2011
Territorial governance	Rural development	Marsden, 1999, Van der Ploeg and Roep, 2003
	Metropolitan areas	Jongeneel et al., 2008; Præstholm & Kristensen, 2007;
	Urban rural relations	Fanfani, 2006; Marsden, 1999; Sabaté Bel & Benito del Pozo, 2010
	Territorial marketing	Fanfani, 2006
Policymaking	Rural development policies	Marsden, 1999; Van der Ploeg and Roep, 2003
	Food policies	Moragues-Faus & Sonnino, 2019; Moragues-Faus & Morgan, 2015
	Farmland preservation policies	Brinkley, 2012; Dwyer, 2011; Perrin et al., 2020; Winter, 2009
	Translocal networks	Moragues-Faus & Sonnino, 2019; Moragues-Faus & Morgan, 2015

Source: Elaborated by the author

References

Abler, D. (2004). Multifunctionality, agricultural policy, and environmental policy. *Agricultural and Resource Economics Review*, *33*(1), 8–17. Scopus. https://doi.org/10.1017/S1068280500005591

Aguglia, L., Henke, R., Poppe, K., Roest, A., & Salvioni, C. (2009). Diversification and multifunctionality in Italy and the Netherlands: A comparative analysis. In *Second Meeting Wye City Group Conference*.

Akram-Lodhi, A. H. (2021). The ties that bind? Agroecology and the agrarian question in the twenty-first century. *The Journal of Peasant Studies*, *48*(4), 687–714. https://doi.org/10.1080/03066150.2021.1923010

Altieri, M. A., & Toledo, V. M. (2011). The agroecological revolution in Latin America: Rescuing nature, ensuring food sovereignty and empowering peasants. *The Journal of Peasant Studies*, *38*(3), 587–612. https://doi.org/10.1080/03066150.2011.582947

Belletti, G., Brunori, G., Marescotti, A., & Rossi, A. (2003). *Multifunctionality and rural development: A multilevel approach.*

Born, B., & Purcell, M. (2006). Avoiding the Local Trap: Scale and Food Systems in Planning Research. *Journal of Planning Education and Research*, *26*(2), 195–207. https://doi.org/10.1177/0739456X06291389

Borras, S. M. (2023). Politically engaged, pluralist and internationalist: Critical agrarian studies today. *The Journal of Peasant Studies*, *50*(2), 449–489. https://doi.org/10.1080/03066150.2022.2163164

Brandt, J., & Vejre, H. (2004). Multifunctional landscapes—Motives, concepts and perceptions. In J. Brandt & H. Vejre (Eds.), *Multifunctional Landscapes* (pp. 3–32). WIT Press.

Brinkley, C. (2012). Evaluating the Benefits of Peri-Urban Agriculture. *Journal of Planning Literature*, *27*(3), 259–269. https://doi.org/10.1177/0885412211435172

Brunori, G., & Rossi, A. (2000). Synergy and Coherence through Collective Action: Some Insights from Wine Routes in Tuscany. *Sociologia Ruralis*, *40*(4), 409–423. https://doi.org/10.1111/1467-9523.00157

Buciega, A., Pitarch, M.-D., & Esparcia, J. (2009). The context of rural-urban relationships in Finland, France, Hungary, the Netherlands and Spain. *Journal of Environmental Policy and Planning*, *11*(1), 9–27. Scopus. https://doi.org/10.1080/15239080902774929

Buijs, A. E., Pedroli, B., & Luginbühl, Y. (2006). From hiking through farmland to farming in a leisure landscape: Changing social perceptions of the European landscape. *Landscape Ecology*, *21*(3 SPEC. ISS.), 375–389. Scopus. https://doi.org/10.1007/s10980-005-5223-2

Busck, A. G., Kristensen, S. P., Præstholm, S., & Primdahl, J. (2008). Porous landscapes – The case of Greater Copenhagen. *Urban Forestry & Urban Greening*, *7*(3), 145–156. https://doi.org/10.1016/j.ufug.2007.05.002

Chiffoleau, Y. (2009). From politics to co-operation: The dynamics of embeddedness in alternative food supply chains. *Sociologia Ruralis, 49*(3), 218–235. Scopus. https://doi.org/10.1111/j.1467-9523.2009.00491.x

Corsi, S., Dell'Ovo, M., Dezio, C., Longo, A., & Oppio, A. (2023). Beyond food: Framing ecosystem services value in peri-urban farming in the post-Covid era with a multidimensional perspective. The case of Cascina Biblioteca in Milan (Italy). *Cities, 137*, 104332. https://doi.org/10.1016/j.cities.2023.104332

De Roest, K. (2000). *The Production of Parimigiano-Reggiano Cheese. The Force of an Artisanal System in an Industrialised World*. Van Gorcum.

Dwyer, J. (2011). UK Land Use Futures: Policy influence and challenges for the coming decades. *Land Use Policy, 28*(4), 674–683. https://doi.org/10.1016/j.landusepol.2010.12.002

Fanfani, D. (2006). Il governo del territorio e del paesaggio rurale nello spazio "terzo" periurbano. Il parco agricolo come strumento di politiche e di progetto. *Ri-Vista. Research for Landscape Architecture, 6*(2), Article 2. https://doi.org/10.13128/RV-17443

Federici, S., & Haiven, M. (2009). On capitalism, colonialism, women and food politics. *Politics and Culture*, (2).

Forman, R. T. T., & Godron, M. (1986). *Landscape Ecology*. Wiley.

Freibauer, A., Rounsevell, M. D. A., Smith, P., & Verhagen, J. (2004). Carbon sequestration in the agricultural soils of Europe. *Geoderma, 122*(1), 1–23. Scopus. https://doi.org/10.1016/j.geoderma.2004.01.021

Gallent, N. (2006). The Rural–Urban fringe: A new priority for planning policy? *Planning Practice & Research, 21*(3), 383–393. https://doi.org/10.1080/02697450601090872

Gentili, A., & Zampetti, G. (Eds.). (2021). *Agroecologia Circolare. Dal campo alla tavola. Coltivare biodiversità e innovazione*. Edizioni Ambiente.

Giunta, I. (2023). Cibo, studi agrari critici ed ecologia politica. In L. Pellizzoni (Ed.), *Introduzione all'ecologia politica* (pp. 331–350). Il Mulino.

Goodman, D., DuPuis, E. M., & Goodman, M. K. (2013). *Alternative Food Networks: Knowledge, Practice, and Politics*. Routledge. https://doi.org/10.4324/9780203804520

Goodman, D., & Watts, M. (Eds.). (1997). *Globalising Food: Agrarian Questions and Global Restructuring*. Routledge.

Guzmán, G., Molina, M., & Guzmán, E. (1999). *Introducción a la Agroecología como Desarrollo Rural Sostenible*. Mundiprensa.

Haase, D., & Nuissl, H. (2007). Does urban sprawl drive changes in the water balance and policy?. The case of Leipzig (Germany) 1870–2003. *Landscape and Urban Planning, 80*(1–2), 1–13. Scopus. https://doi.org/10.1016/j.landurbplan.2006.03.011

Hägerstrand, T. (1995). *A Look at the Political Geography of Environmental Management*. Department of Geography, University College Dublin.

Hall, C., McVittie, A., & Moran, D. (2004). What does the public want from agriculture and the countryside? A review of evidence and methods. *Journal of Rural Studies, 20*(2), 211–225. Scopus. https://doi.org/10.1016/j.jrurstud.2003.08.004

Holloway, L., Kneafsey, M., Venn, L., Cox, R., Dowler, E., & Tuomainen, H. (2007). Possible Food Economies: A Methodological Framework for Exploring Food Production–Consumption Relationships. *Sociologia Ruralis, 47*(1), 1–19. https://doi.org/10.1111/j.1467-9523.2007.00427.x

Holt Giménez, E., & Shattuck, A. (2011). Food crises, food regimes and food movements: Rumblings of reform or tides of transformation? *The Journal of Peasant Studies, 38*(1), 109–144. https://doi.org/10.1080/03066150.2010.538578

Horst, M., McClintock, N., & Hoey, L. (2017). The Intersection of Planning, Urban Agriculture, and Food Justice: A Review of the Literature. *Journal of the American Planning Association, 83*(3), 277–295. https://doi.org/10.1080/01944363.2017.1322914

Jansma, J. E., Veen, E. J., & Müller, D. (2024). Beyond urban farm and community garden, a new typology of urban and peri-urban agriculture in Europe. *Urban Agriculture & Regional Food Systems, 9*(1), e20056. https://doi.org/10.1002/uar2.20056

Jarosz, L. (2008). The city in the country: Growing alternative food networks in Metropolitan areas. *Journal of Rural Studies, 24*(3), 231–244. https://doi.org/10.1016/j.jrurstud.2007.10.002

Jongeneel, R. A., Polman, N. B. P., & Slangen, L. H. G. (2008). Why are Dutch farmers going multifunctional? *Land Use Policy, 25*(1), 81–94. Scopus. https://doi.org/10.1016/j.landusepol.2007.03.001

Knickel, K. (1999). Agricultural landscapes as places of production and for living in owner's versus producer's decision making and the implications for planning. *Landscape and Urban Planning, 46*(1–3, 143), –150. https://doi.org/10.1016/S0169-2046(99)00038-9

Knickel, K., & Renting, H. (2000). Methodological and Conceptual Issues in the Study of Multifunctionality and Rural Development. *Sociologia Ruralis, 40*(4), 512–528. https://doi.org/10.1111/1467-9523.00164

Lamptey, B. L., Barron, E. J., & Pollard, D. (2005). Impacts of agriculture and urbanization on the climate of the Northeastern United States. *Global and Planetary Change, 49*(3–4), 203–221. Scopus. https://doi.org/10.1016/j.gloplacha.2005.10.001

Lohrberg, F., Christenn, K., Timpe, A., & Sancar, A. (2022). *Urban Agricultural Heritage*. Birkhäuser. https://doi.org/10.1515/9783035622522

Lohrberg, F., Licka, L., Scazzosi, L., & Timpe, A. (Eds.). (2015). *Urban Agriculture Europe*. Jovis.

Lowe, P., Murdoch, J., & Ward, N. (1995). Beyond endogenous and exogenous models: Networks in rural development. In J. D. van der Ploeg & G. van Dijk (Eds.), *Beyond Modernization: The Impact of Endogenous Rural Development* (pp. 87–105). Van Gorcum.

Moragues-Faus, A., & Morgan, K. (2015). Reframing the foodscape: The emergent world of urban food policy. *Environment and Planning A: Economy and Space, 47*(7), 1558–1573. https://doi.org/10.1177/0308518X15595754

Moragues-Faus, A., & Sonnino, R. (2019). Re-assembling sustainable food cities: An exploration of translocal governance and its multiple agencies. *Urban Studies, 56*(4), 778–794. https://doi.org/10.1177/0042098018763038

Paül, V., & McKenzie, F. H. (2013). Peri-urban farmland conservation and development of alternative food networks: Insights from a case-study area in metropolitan Barcelona (Catalonia, Spain). *Land Use Policy, 30*(1), 94–105. https://doi.org/10.1016/j.landusepol.2012.02.009

Perrin, C., Clément, C., Melot, R., & Nougarèdes, B. (2020). Preserving Farmland on the Urban Fringe: A Literature Review on Land Policies in Developed Countries. *Land, 9*(7), 223. https://doi.org/10.3390/land9070223

Piorr, A., Zasada, I., Doernberg, A., Zoll, F., & Ramme, W. (2018). *Research for AGRI Committee–Urban and Peri-urban Agriculture in the EU.* Policy Department for Structural and Cohesion Policies.

Præstholm, S., & Kristensen, S. (2007). Farmers as Initiators and Farms as Attractors for Non agricultural Economic Activities in Peri-urban Areas in Denmark. *Danish Journal of Geography, 107*, 13–27. https://doi.org/10.1080/00167223.2007.10649566

Renting, H., Rossing, W., Groot, J., Ploeg, J., Laurent, C., Perraud, D., Stobbelaar, D., & Ittersum, M. K. (2009). Exploring multifunctional agriculture. A review of conceptual approaches and prospects for an integrative transitional framework. *Journal of Environmental Management, 90*(Suppl. 2), S112–S123. https://doi.org/10.1016/j.jenvman.2008.11.014

Robinson, G. (2003). *Geographies of Agriculture. Globalisation, Restructuring and Sustainability.* Routledge. https://doi.org/10.4324/9781315839509

Sabaté Bel, J., & Benito del Pozo, P. (2010). Paisajes culturales y proyecto territorial: Un balance de treinta años de experiencia. *Identidades territorio cultura patrimonio, 2*, 2–21. https://doi.org/10.5821/identidades.8786

Sonnino, R., & Marsden, T. (2006). Beyond the divide: Rethinking relationships between alternative and conventional food networks in Europe. *Journal of Economic Geography, 6*(2), 181–199. https://doi.org/10.1093/jeg/lbi006

Susi Botto, I., & Di Vita, S. (2017). *Oltre l'Expo 2015. Tra dimensione ordinaria e straordinaria delle politiche urbane.*

Tornaghi, C. (2014). Critical geography of urban agriculture. *Progress in Human Geography, 38*(4), 551–567. https://doi.org/10.1177/0309132513512542

Tornaghi, C., & Dehaene, M. (2020). The prefigurative power of urban political agroecology: Rethinking the urbanisms of agroecological transitions for food system transformation. *Agroecology and Sustainable Food Systems, 44*(5), 594–610. https://doi.org/10.1080/21683565.2019.1680593

Van Broekhuizen, R., Klep, L., Oostindie, H., & van der Ploeg, J. D. (Eds.) (1997). *Renewing the countryside, an atlas with two hundred examples from Dutch rural society*. Misset.

Van Depoele, L. (1996). European rural development policy. In H. Heijman, H. Hetsen, & J. Frouws (Eds.), *Rural reconstruction in a market economy*. Backhuys Publishers.

Van Der Ploeg, J., & Roep, D. (2003). Multifunctionality and rural development: The actual situation in Europe. In *Multifunctional Agriculture; A newparadigm for European Agriculture and Rural Development* (pp. 37–53). Ashgate.

Van Huylenbroeck, G., Vandermeulen, V., Mettepenningen, E., & Verspecht, A. (2007). Multifunctionality of Agriculture: A Review of Definitions, Evidence and Instruments. *Living Reviews in Landscape Research, 1*. https://doi.org/10.12942/lrlr-2007-3

Vandermeulen, V., Verspecht, A., Van Huylenbroeck, G., Meert, H., Boulanger, A., & Van Hecke, E. (2006). The importance of the institutional environment on multifunctional farming systems in the peri-urban area of Brussels. *Land Use Policy, 23*(4), 486–501. Scopus. https://doi.org/10.1016/j.landusepol.2005.06.002

Vejre, H., Abildtrup, J., Andersen, E., Andersen, P. S., Brandt, J., Busck, A., Dalgaard, T., Hasler, B., Huusom, H., Kristensen, L. S., Kristensen, S. P., & Præstholm, S. (2007). Multifunctional agriculture and multifunctional landscapes—Land use as an interface. In Ü. Mander, H. Wiggering, & K. Helming (Eds.), *Multifunctional Land Use: Meeting Future Demands for Landscape Goods and Services* (pp. 93–104). Springer. https://doi.org/10.1007/978-3-540-36763-5_6

Ward, N. (1993). The Agricultural Treadmill and the Rural Environment in the Post-Productivist Era. *Sociologia Ruralis, 33*(3–4), 348–364. https://doi.org/10.1111/j.1467-9523.1993.tb00969.x

Wascher, D., Schafer, L., Eyre, P., Henriksen, C. B., & Parente, S. (2023, June). *Long-term living lab support towards food system transformation. Policy recommendations for long-term support of Living Labs contributing to food system transformation. A joint initiative from the EU FOOD 2030 Project Family*. European Commission.

Wezel, A., Bellon, S., Doré, T., Francis, C., Vallod, D., & David, C. (2009). Agroecology as a science, a movement and a practice. A review. *Agronomy for Sustainable Development, 29*(4), 503–515. https://doi.org/10.1051/agro/2009004

Whatmore, S., & Thorne, L. (2004). Nourishing Networks: Alternative Geographies of Food. In *Reading Economic Geography* (pp. 235–248). John Wiley & Sons. https://doi.org/10.1002/9780470755716.ch15

Winter, M. (2009). Agricultural land use in the era of climate change: The challenge of finding 'Fit for Purpose' data. *Land Use Policy, Land Use Futures, 26*, S217–S221. https://doi.org/10.1016/j.landusepol.2009.08.028

Zasada, I. (2011). Multifunctional peri-urban agriculture—A review of societal demands and the provision of goods and services by farming. *Land Use Policy, 28*(4), 639–648. https://doi.org/10.1016/j.landusepol.2011.01.008

CHAPTER 3

Planning, Design, and Landscape Management Mechanisms

Introduction

What brings these three European case studies together, and what played an important role in their selection, is their shared lineage in addressing the challenges of recognizing areas of food production and agricultural activity. Their selection was based on a research process that began by identifying planning and design mechanisms that place agricultural production at the center of their agenda. Within this search, the *agricultural park* emerged as a specialized yet relatively under-disseminated mechanism.

The selection focused on identifying territories where this mechanism has been implemented, remains active, and where its current analysis can reveal how it has evolved and how it addresses contemporary challenges. These cases are examined from a metropolitan perspective, which emerges not only as a position adopted by the research but as a necessary scale for establishing dialogue between these mechanisms, territories, and several of the arguments discussed in Chap. 2. In this way, this chapter introduces the three territories—Milan, Barcelona, and the Ruhr Metropolis—through their respective planning and landscape management instruments, employing a narrative that acknowledges the dialogue between these contexts and some of the complementary strategies that reinforce a shared pursuit of protection, dynamization, and management of the agrarian landscape and the elements that structure it. This chapter thus acts as

© The Author(s), under exclusive license to Springer Nature Switzerland AG 2026
A. J. Salvador, *Metropolitan Food Systems*, Sustainable Urban Futures, https://doi.org/10.1007/978-3-032-26016-1_3

a premise for understanding some of the decisions and the framework within which these cases will be primarily analyzed throughout this book.

Rooted in the rise of industrial society and the modern city, the urban-rural and nature-society dichotomies have long represented some of the most challenging relationships in planning and governance. These dichotomies denote an asymmetrical relationship in which planning and research practices have predominantly overlooked the rural dimension, often treating it "as an extension of urban planning and urban studies reapplied in contexts beyond the metropolis" (Gkartzios et al., 2022, p. 3). At the same time, planning and urban design have sought to challenge these divides through instruments developed to mitigate urban sprawl during the twentieth century and to address the formation of fringe belts in major urban centers (Lemes de Oliveira, 2019; Whitehand & Morton, 2006). Mechanisms such as the Green Belt, first implemented in British cities and later adopted across Europe, stand as significant examples, alongside approaches embedded in housing typologies, most notably the Garden City Movement (Keil & Macdonald, 2016; Lemes de Oliveira, 2019; Zepp, 2018).

Although the *green* component of these instruments has often been addressed in a generic manner (Gallent, 2006), other approaches have sought to engage more directly with the diverse ecologies of peri-urban belts, particularly their productive and agricultural dimensions and their proximity to cities (Ferraresi & Rossi, 1993; Rettig, 1976; Timpe, 2020). The *ceinture verte* in Rennes and the agricultural park (*parco agricolo*) in Milan are among the most prominent examples, with the latter serving as the point of departure for this chapter. From a landscape and spatial planning perspective, the range and scale of agricultural interventions have expanded, reinforced by landscape management practices and active forms of citizenship as tools for placemaking (Donadieu, 2013; Mininni, 2013; Verzone & Woods, 2021). Within this context, the figure of the *park*, or defining protected land, remains a recurring strategy for peri-urban natural areas, framed as a multifunctional device providing leisure, environmental, and social services, and integrated into broader green infrastructure systems.

The conception of the *agricultural park* emerged in the 1980s, with the first widely recognized examples implemented during the 1990s, notably Parco Agricolo Sud in Milan, Parco Agricolo dei Ciaculli in Palermo, and Parc Agrari del Baix Llobregat in Barcelona (Yacamán Ochoa & Zazo Moratalla, 2015). Since then, related definitions of multifunctional agricultural parks have developed primarily within the European context—especially in Spain, Italy, and France—as well as in Germany through the concept of *produktive parks* (Fanfani, 2006, 2019; Poli, 2015; Timpe, 2017; Yacamán Ochoa & Zazo Moratalla, 2015; Zazo Moratalla & Paul, 2022). Beyond Europe, experiences in countries such as Argentina have been directly influenced by these Spanish precedents (Esteves, 2022; Mantovani et al., 2020).

Despite variations in scale and governance structures, agricultural parks have been primarily framed as strategies for farmland preservation and urban containment (Jarrige & Perrin, 2017; Zasada, 2011). As noted by Yacamán Ochoa and Zazo Moratalla (2015, p. 14), early agricultural parks developed in the 1990s and early 2000s functioned as mechanisms of "territorial defense" during periods of intense real estate expansion, whereas those planned in the following decade increasingly focused on agricultural profitability within a context of global structural crisis. More recently, attention has expanded toward broader concerns, including territorial cohesion and sustainability, food sovereignty, local development, and the strengthening of food supply chains.

From this perspective, agricultural activities are understood as cultural and educational services that foster new interactions between the city and the countryside. This model encourages farms and farmers to reestablish direct relationships with urban areas through food production, the diversification of activities, and the generation of additional income derived from social engagement (Magnaghi, 2000). The rise of the environmental agenda at the beginning of the century, together with ongoing crises, has revitalized the discourse on the peri-urban space as a green, multifunctional infrastructure for cities and their inhabitants. This space is no longer conceived solely as a large, continuous green space for recreation and education, but also as an environmental resource capable of assuming productive roles in food production, energy generation, and waste processing (Gallent, 2006; Vejre et al., 2007).

The concept of *multifunctionality* encompasses economic, ecological, community, and aesthetic dimensions (Brandt & Vejre, 2004; 2000; Vejre et al., 2007) and should not be understood merely as a set of post-industrial activities offered by farms in peri-urban and rural areas close to cities (van der Ploeg et al., 2009). As Gallent (2006) argues, approaches to planning in peri-urban areas have shifted since the turn of the century from a "land-use model" toward a "broader spatial agenda," in which spatial planning is tasked with assigning functions and future forms of land use, while acting as a facilitator of "negotiated governance" (Gallent, 2006, p. 389).

The Urban Countryside

Pierre Donadieu's *campagnes urbaines* (urban countryside), conceived as a "realistic utopia," is one of the foundational theoretical frameworks in Europe that repositions peri-urban agricultural areas as assets for the creation of hybrid landscapes. Rather than treating the peri-urban as a residual fringe, this concept challenges the effects of urban sprawl by reframing these peripheral territories—often neglected by planning and rural land-use policies—as spaces capable of restoring relationships between inhabitants and their environment (Donadieu, 2013). Defined as an "oxymoron with a planning intention," the urban countryside is not merely interpretative but prescriptive, aiming to actively shape these territories through landscape projects centered on place resignification (Donadieu, 2013, p. 74).

Within this framework, agricultural land and woodlands are conceived as structuring elements of urban form and as a natural infrastructure of public interest. Donadieu identifies three planning instruments: the Landscape Plan, the Landscape Charter, and the Countryside Park. The latter envisioned as an agricultural and forestry archipelago embedded within the urban fabric, encompassing open spaces around cities rather than clearly bounded areas. These parks integrate diverse forms of agricultural production, from hobbyist to peri-urban farming, generating social and educational value through direct interaction between urban populations and productive landscapes. Donadieu's approach underscores the central role of landscape planning, political commitment, and citizen participation, illustrated by French examples such as Perigny-sur-Yerres, Parc des Lilas in Vitry-sur-Seine, and Parc de la Lironde in Montpellier.

An Agricultural Green Belt in Milan: The First Agricultural Park and the Emergence of a New Typology

The idea of creating a large agricultural belt for Milan, inspired by British and European green belt examples, emerged in the 1967 and 1975 *Piani Intercomunali* (Balducci & Piazza, 1981; Ferraresi & Rossi, 1993), led by

Milan's southern municipalities to support agriculture and protect farmland. The National Institute of Urbanism (INU) promoted this approach, emphasizing agriculture's productive role near the city (Balducci & Piazza, 1981). The concept was formally recognized in 1987 through a citizens' initiative law (*legge di iniziativa popolare*), which became a regional law in 1990.

Balducci and Piazza (1981) note that the Piano Intercomunale Milanese (PIM) aimed to coordinate metropolitan planning across Milan and surrounding municipalities, anticipating greater demographic growth than occurred. This led to real estate liberalization, with retail centers, private health facilities, office parks, and large residential developments for upper-income groups. The resulting oversupply, farmland loss, and land speculation reshaped land ownership and the real estate market. These developments, marketed through campaigns resembling mid-century North American strategies, were supported by flexible urban planning regulations and produced homogeneous, disaggregated urban forms inserted into the countryside, near new leisure and commercial hubs (Balducci & Piazza, 1981; Salvador, 2025). These changes transformed the environment and local lifestyles, as smaller municipalities linked development to economic growth and urbanization fees (*oneri di urbanizzazione*).

Ferraresi and Rossi (1993), in *Il Parco come cura e coltura del territorio*, document Parco Agricolo Sud's design process and defines its theoretical framework. The publication captures a critical moment in Milan, situating the emergence of the agricultural park as a new planning and design typology. The authors drew inspiration from Giovanni Astengo's work in Piemonte and Assisi, green policies and open space systems in Bologna, London, and Erlangen, and historical parks such as Giardini di Arezzo and Parco di Ferrara, known as Il Barco. While the Arezzo project was not integrated into the city's Development Plan, it remains a notable case where agriculture and the agrarian landscape informed territorial planning, particularly the Territorial Plan (*Piano Territoriale di Coordinamento—PTC*) coordinated by Gian Franco Di Pietro in the early 2000s, whose work in Siena and sensitivity to the agrarian landscape has been recognized in the Italian context as a major contribution to planning (Poli, 2023).

Parco Agricolo Sud's design process also included formal exchanges with emerging agricultural parks in Palermo and Barcelona through a 1994–1998 Life EU project. However, Milan's case remains exceptional in scale (47,000 hectares), compared with Parco Agricolo dei Ciaculli in Palermo (approximately 800 hectares) and Parc Agrari del Baix Llobregat in Barcelona (initially of 2938 hectares).

At the end of the twentieth century, Milan was heavily urbanized and intensively farmed (Balducci & Piazza, 1981; Ferraresi & Rossi, 1993). The concept of *fare parco* ("make a park") aimed to emphasize an awareness of limits—not only as boundaries but as the finite nature of resources—seeking to halt consumption, preserve, reclaim, and restore the territory. The park was conceived to recover the territorial complexity lost under intensive agriculture that no longer shaped the landscape (Ferraresi & Rossi, 1993). Its introduction was motivated by its protective and restrictive role, environmental compatibility with peri-urban morphology, and its capacity to host agricultural, educational, and ecological activities (Ferraresi & Rossi, 1993).

Currently, Parco Agricolo Sud has successfully preserved agricultural land amid urban expansion, yet it remains politically contested among the 61 municipalities within its 47,000 hectares (see Fig. 3.1). As noted by Vescovi (2012), the park has struggled to develop a strategic, city- and park-wide vision, limiting its capacity to lead projects and foster networks (Salvador, 2024b). These gaps have prompted non-institutional initiatives, mainly by farmers, to adapt production to changing market demands. Rural agricultural districts (*distretti agricoli rurali*) now coordinate collective strategies, reduce costs, and explore less intensive production, while solidarity purchasing groups (GAS)[1] connect consumers to producers, shorten supply chains, and stabilize farmer income (Calori & Magarini, 2015; Vescovi, 2012). The Milanese case thus served as both a testing ground for a new agricultural park typology and an ad-hoc mechanism for addressing the complex challenges of peri-urban agrarian spaces.

The Multifunctional Agricultural Park within the Territorialist School of Thought

The contribution of the *Scuola dei Territorialisti*, led by Alberto Magnaghi and a team of researchers from the University of Florence, has been crucial in establishing the theoretical framework for the agricultural park and multifunctional landscapes. Their work, influenced by Lewis Mumford

[1] The Italian name for Community Supported Agriculture (CSA) initiatives, which will be further addressed by Chap. 8.

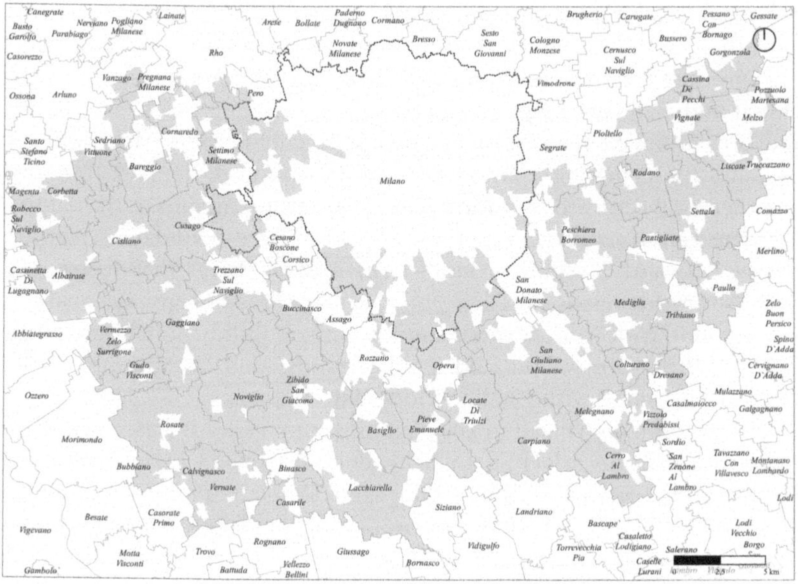

Fig. 3.1 Municipalities inside Parco Agricolo Sud Milano. (Source: Elaborated by the author)

and the concept of the "urban bioregion," integrates agricultural territories and food production into a polycentric and metabolic relationship between urban settlements and landscape features (Fanfani & Matarán Ruiz, 2020; Magnaghi & Fanfani, 2010). In Central Tuscany, multifunctional agricultural parks are conceived as a *pact* between city and countryside, acknowledging the complexity and specificities of the territory and fostering "place consciousness" based on local identity (Magnaghi, 2000). Magnaghi critiques modernist urbanization for *deterritorializing* society, proposing the bioregion model as an alternative system of cities, places, and natural elements, emphasizing the inseparability of nature and culture, and the need to rebuild relationships between *rurality* and *urbanity* (Magnaghi, 2011).

The 2010 publication *Patto città–campagna: un progetto di bioregione urbana per la Toscana centrale* frames the agricultural park as both a process and a project: a bottom-up social process and a multidisciplinary, multisectoral institutional project (Magnaghi & Fanfani, 2010). Multifunctionality encompasses strategic functions that support an "urban ecosystem" linking urban open spaces and peri-urban agricultural areas, including hydric resources, waste management, leisure, and education. The proposed *green core* along the Arno River draws inspiration from Dutch and Belgian planning experiences, integrating cities, agriculture, and key natural elements into a landscape-oriented vision. Magnaghi and Fanfani (2010) highlight an operational dimension of the park, bringing together local actors to transform open-space planning into territorial spatial plans (*piani territoriali*) and moving from fixed boundaries (*perimetrazione*) toward "testing grounds" (*area laboratorio*) that foster new city-countryside relations. Their *project* acts in three dimensions: enabling a city-countryside pact, experimenting with multifunctional agriculture that produces both food and public goods, and integrating urban-territorial planning with rural development.

This vision materialized in key planning commissions: Tuscany's *Piano di Indirizzo Territoriale* (PIT) in 2014, Prato Province's *Piano Territoriale di Coordinamento* (PTC) in 2003, and Puglia's *Piano Paesaggistico Territoriale* (PPTR) in 2010, incorporating landscape-oriented governance into regulatory plans. The 2003 Prato PTC transformed the agricultural park into a territorial strategy to restore lost natural and agricultural attributes, creating synergies between city and countryside through planning, design, and policy (Fanfani, 2009). The 7000-hectare Prato case involved eight municipalities and engaged multiple actors through design charrettes, producing a masterplan aimed at redirecting agriculture toward multifunctional models and local food demands. Initiatives included short supply chains for *Calvana* bovine meat and cereals, collaborations with local food procurement programs, and the promotion of agroecology alongside non-agricultural services. The Prato experience underscores the importance of integrating planning and policy, fostering shared visions among public authorities, private actors, farmers, and citizens, and sustaining bottom-up initiatives while embedding them within a broader territorial strategy (Fanfani, 2009).

The Consolidation of the Spanish Model: The *Parque Agrario*

Spain is a country where agricultural parks have significantly developed as planning and governance mechanisms, promoted by farmers' associations, municipalities, supra-local governments, academia, and civil society.

Spanish scholars have contributed to defining a model for the agricultural park (*parque agrario*). Ana Zazo's doctoral thesis (2015) and the book *El Parque Agrario: una figura de transición hacia nuevos modelos de gobernanza territorial y alimentaria*, curated by Carolina Yacamán Ochoa and Ana Zazo Moratalla (2015), represent one of the most comprehensive national analyses of agricultural parks. The publication examines multiple cases—implemented or not—highlighting the diverse pathways and planning mechanisms used to design such parks.

Joaquin Sabaté Bel's (2015) analysis of the Baix Llobregat case underlines the need to structure parks through adaptable planning and project mechanisms capable of responding to evolving demands and resisting transformation processes. He emphasizes understanding rural areas beyond a simple *agricultural label*, recognizing the spatial fabric—including morphology, plots, roads, waterways, and built structures—and integrating design within complex planning and management frameworks. Sabaté Bel also highlights the detrimental impact of mobility and logistics decisions on rural and agrarian dynamics. Along this book, the Baix Llobregat case will be further discussed within the broader system of agricultural parks currently coordinated at the provincial level in Barcelona, continuing the efforts initiated by Josep Montasell in the Maresme area since the 1980s (Paül, 2015).

The Baix Llobregat Experience as a Key Instance for Defining the Character of Agricultural Parks

Analysis of planning documents, beginning with the 1996 study *Estudi sobre la viabilitat d'implementació d'un parc agrícola a la comarca del Baix Llobregat* by the Diputació de Barcelona and the local farmer's association, Unió de Pagesos de Catalunya, reveals an interdisciplinary attempt to conceive an agricultural park for the Baix Llobregat county (*comarca*). The proposal highlighted how urbanism's limited understanding of agrarian territories had diminished agriculture as a land use. Previous initiatives, such as the 1976 General Metropolitan Plan and the 1992 Natural Areas of Special Interest Plan (PEIN), were generic or excluded agriculture, while the 1993–2007 Infrastructure Strategic Plan (PDI) introduced projects that significantly impacted farmland and landscape integrity. The Baix Llobregat proposal addressed fragmented territorial governance through three central components: a management entity coordinating planning, governance, and economic-productive needs; a land-use plan to protect and regulate multiple functions and define a park perimeter; and a development plan integrating social, economic, and environmental assets.

The first proposal by CCRS arquitectos (1997–1998), *Pla Especial del Parc Agrari del Baix Llobregat—Avanç de Pla*, emphasized creating a "new identity" for metropolitan agricultural areas. Unlike Milan's Parco Agricolo Sud, which prioritizes environmental and landscape dimensions, the Baix Llobregat plan focused on production while balancing ecological and social values, particularly regarding the Llobregat River and its Delta. The 1998 Special Plan analyzed other European parks at local, supra-local, and provincial scales, classifying land use as cultural, agricultural, or natural, and park focus as social, ecological, or productive. Examples included Duisburg Nord Landscape Park (Germany, 200 ha) and Parco Agricolo dei Ciaculli (Palermo, 800 ha) for local parks; Parco delle Groane (Milan, 3400 ha, 16 municipalities) for supra-local; and Parco Agricolo Sud (46,308 ha, 61 municipalities) for provincial. This analysis informed the design of a management entity suitable for coordinating nearly 4000 hectares across 13 municipalities, emphasizing the link between park size and its political-administrative structure. The proposal also acknowledged the need to integrate the agricultural park within Barcelona's green system and produce a larger and integrated territorial unit. Some of the examples cited are London's *Green Belt*, Ottawa's *Ceinture de Verdure*, the Dutch *Groene Hart*, Paris' *Ceinture Verte*, and the *Banlieue* and Valencia's *Pla Verd*.

As Sabaté Bel (2000) explains, redesigning the Baix Llobregat territory relied on three main elements: its formal structure, its spatial qualities, and its historical formation. The 1998 Special Plan (*pla especial*) supports this conceptual approach, recognizing the Llobregat River not only as a natural resource but also as a transformative formal element. The river and the nineteenth-century canal system shaped the Delta into a large irrigated plain capable of supplying Barcelona with fresh produce (Rocamonde & Sabaté Bel, 2019). This historical understanding of its evolution establishes a historical socio-spatial relationship. According to Sabaté Bel (2000), the agrarian structure and spatial qualities involve acknowledging the basic elements such as soil, irrigation network, roads, and buildings, which compose the tissue that structures and enables the production of these agricultural mosaics. Therefore, the Baix Llobregat plain is a heterogeneous space, composed of agrarian units shaped by parcel size, access to water, proximity to the sea or surrounding hills, land ownership patterns, relationships with nearby towns, and crop types, and further structured by the social cohesion fostered through cooperatives and local associations. More recently, mobility infrastructure has severely altered both the landscape and farm dynamics.

Towards a Spanish Model for Agricultural Parks

Zazo Moratalla and Paül (2022) provide a comprehensive overview of Spain's attempts to structure agricultural parks. Of 19 identified processes, only 4 have been implemented: Sabadell and Baix Llobregat in Barcelona's metropolitan area, and Rivas and Fuenlabrada (Soto del Grillo) in Madrid's metropolitan area. The Madrid parks, created in the 2010s, are smaller and less administratively complex than Baix Llobregat. Even when formal constitutions were not achieved, many processes generated significant social and political engagement, such as the Cinc Sènies park in Maresme, the Delta de la Tordera, and the Conca d'Òdena, now coordinated at the provincial level in Barcelona (Paül, 2015).

Baix Llobregat's park benefited from a robust political and technical process, with a management entity of 10 staff and a full-time manager overseeing daily operations, communication, events, farmer relations, and municipal coordination. This model, however, proved difficult to replicate. Newer initiatives in the 2020s, such as Espai Agrari Baixa Tordera, involve local consortia supported technically by the Diputació de Barcelona, guided by a Management and Development Plan rather than a full Territorial and Spatial Plan (Salvador, 2024a). These projects address the specific challenges and production needs of peri-urban agriculture, often facilitating shared infrastructure and generational transitions, as seen in the Parc Rural del Montserrat, through land banks (*banc de terres*) and shared slaughterhouses, known as *obradors compartits*.

In Madrid, the Fuenlabrada and Soto del Grillo parks pursue agroecological models while responding to local dietary needs, despite limited irrigation and urban pressures. Fuenlabrada (800 ha, 220 ha irrigable) relies on small, family-run farms (*minifundios*) and initiatives *like the Cómete Fuenlabrada* direct sales program to strengthen local markets and economic sustainability (Olmo & Yacamán Ochoa, 2015). Soto del Grillo (90 ha municipal farm) uses a land bank to allocate plots, shared tools, irrigation access, and short supply chains to schools, markets, and consumers. Both Madrid parks illustrate a *slender* model of agricultural park: smaller in scale and governance but driven by local municipal commitment and strategic policies that reestablish links between city and agricultural areas.

Towards Planning Instruments and Design Tools for Agrarian Spaces

Although initial proposals, such as Ferraresi and Rossi (1993) for Milan, the Baix Llobregat design process by (Sabaté Bel, 2015), and the Territorialists' initiatives in Tuscany (Fanfani, 2009; Magnaghi & Fanfani, 2010) demonstrate a strong design component, this aspect is often missing in newer projects. Zazo Moratalla and Paül (2022) highlight that a proper and formal plan—crucial for structuring the Baix Llobregat to support agriculture—is absent in recent initiatives, such as the Baixa Tordera and the Madrid parks Fuenlabrada and Soto del Grillo. In Spain, this weak planning presence reflects the supra-municipal scope of agricultural parks, while land-use planning remains primarily a local municipal activity, raising issues of scale, tools, and governance (Zazo Moratalla & Paul, 2022). Catalunya's 2019–3 Regional Law (*Ley 3–2019 de los espacios agrarios*) represents a first step toward enabling smaller municipalities to draft sectorial agricultural plans, offering an opportunity to reintroduce design and land-use components similar to the Baix Llobregat experience.

This gap between early parks, scholarly models, and later developments led to exploring a third context: North-Rhine Westphalia, Germany, where urban design with spatial and landscape approaches has contributed to peri-urban park development. Productive-oriented landscape parks with significant agricultural components continue a design-focused approach, as seen in Duisburg-Nord Landschaftspark, Hugo Biomass Park, Gluckaufpark Hassel (Gelsenkirchen), Aachener Soers Pferdelandpark (Aachen), Grünes C (Bonn), and Belvedere Agricultural Park and Waldlabor Cologne (Cologne) (Matros & Lohrberg, 2016; Timpe, 2019, 2020). The latter includes a 25-hectare experimental short-rotation forest lab and exemplifies integration of landscape management with productive park design. Within this framework, productive parks function as strategic spaces in polycentric urban regions for post-industrial regeneration and co-created green infrastructure, frequently supported by EU-funded projects.

Timpe's (2017) doctoral research extended beyond Germany, analyzing parks in Baix Llobregat (Spain) and Parco Agricolo Sud (Italy), establishing technical and theoretical links that enrich planning and design through a landscape-oriented approach. His approach conceptualizes parks integrating a primary economic activity—agriculture—through three landscape dimensions: as a product of nature and culture, as a social construct, and as an aesthetic construct. In productive parks, these dimensions translate into ecological, social, and aesthetic functions. Unlike Montasell's model, which prioritizes agricultural production and economic value, Timpe emphasizes the multifunctional capacity of the landscape to produce a complex set of outputs, largely agricultural but not exclusively, resonating with Magnaghi's (2000) metabolic understanding and the multifunctional landscape discussion in Chap. 2 (Vejre et al., 2007).

Timpe's work is further enriched by RWTH Aachen University's Institute of Landscape Architecture through the COST Action Urban Agriculture Europe project (see Lohrberg et al., 2015), which adopts a transdisciplinary perspective on urban agriculture. This framework positions spatial planning as capable of engaging with agricultural practice at multiple scales and geographies, recognizing diverse planning tools, forms of engagement, and the spatial and social roles of urban agriculture across Europe.

Productive Parks and Scalable Green Infrastructure in North Rhine—Westphalia

The concept of a *productive park* (Timpe, 2017) links landscape design, spatial planning, and production—both agricultural and energy—as a reintegrated element within planning disciplines. Timpe (2019) situates productive parks within the historical evolution of the park, noting the shift from productivity to leisure as the primary function. In Germany, parks have served as "experimental fields" for designing and testing human-nature relationships, including small-scale experiments later scaled to larger landscapes, echoing eighteenth-century English parks as experimental grounds for agricultural innovations (Timpe, 2019). The contemporary productive park revives pre-industrial forms where production and nature coexist, countering the fossil-dependent, labor-intensive urban model that redefined parks as leisure facilities. Timpe highlights the International Building Exhibition (IBA) Emscher Park as a key context that explored post-industrial aesthetics in combination with wilder landscape forms.

An Ecological Restoration Pathway: The IBA Emscher Landscape Park

To understand the Regionalverband Ruhr's ecological restoration of the Ruhr Metropolis, it is essential to analyze the IBA Emscher Park and its evolution into the Emscher Landscape Park, a system of green belts and open spaces. The Internationale Bauausstellung (IBA) is a historic program originating in the twentieth century, initially to showcase architectural achievements. The 1979–1987 IBA Berlin focused on post-war urban reconstruction, while the 1989–1999 IBA Emscher Park redeveloped former industrial sites at innovative territorial scales, influencing later programs such as IBA Fürst-Pückler-Land (2000–2010) and IBA Stadtumbau (2002–2010). The IBA Emscher Park spanned over 800 km^2, integrating spatial planning, landscape design, and architecture to promote ecological, cultural, and economic transformation. RVR and 17 municipalities implemented 117 projects financed by EU, federal, and regional funds, with a 2.5-billion-euro budget. Key initiatives included establishing a regional park of 300 km^2, restoring the hydric system, developing pedestrian and cycling infrastructure, and preserving industrial archaeological elements as cultural landmarks (Gruehn, 2017).

A central legacy of the IBA was its technical landscape planning process, which assigned municipalities to develop landscape plans following a unified framework, methods, and content (Kipar, 1993). This process led to the master plan for the Emscher Landscape Park (2010), covering 457 km^2, including the Emscher Valley, and outlining over 400 projects for a 30-year implementation period (Gruehn, 2017). Although farmland and agriculture were initially less defined, Timpe (2019) notes that several interventions incorporated productive elements as part of design. For example, the Duisburg-Nord Landscape Park includes an educational farm with social functions, while the Mechtenberg Landscape Park (Bochum, Essen, Gelsenkirchen) integrates farming and agricultural aesthetics, offering direct sales, a cafe, and self-harvesting fields, exemplifying the productive park concept in practice. Multifunctional farming activities play a key role in Timpe's conceptualization of productive parks.

Planning, Design, and Landscape Management: An Integrated Perspective

This chapter has analyzed approaches implemented in various cases towards transcending farmland as a land-use and adopting planning and management strategies and tools that acknowledge the complexity of agrarian territories and embrace their production dimension The agricultural park emerges as a common resource in the these contexts, and also as an element of knowledge exchange, acknowledging it as a planning, management, and governance mechanism for peri-urban agricultural areas. The initial approach to the term recognizes its theoretical foundation, among others embodied in Donadieu's definition of *urban countryside*, which describes the countryside's pressures, challenges, potential, and aesthetics at the edge of cities and emerging amidst urban agglomerations. A carefully delineated narrative was used to analyze the earlier cases and the first attempts to define this emerging, hybrid, and unusual typology. The analysis builds on a rich body of research, predominantly conducted in the Spanish case, to further discuss the possible models that emerge from these scholarly activities and how they differ from the practical models of the few implemented parks. Despite agreement on the parks' objectives in terms of farmland protection, among the crucial aspects that emerge in this analysis are the forms and content of these parks and the types of tools used for their implementation.

Some of the most discussed aspects in the conceptual models of an agricultural park, and during academic debates, are the integration of planning, management, and governance. However, as shown especially by Zazo Moratalla and Paül (2022), implementing agricultural parks appears to be much more challenging than expected. Although agency and political commitment are key, as evidenced in the Barcelona cases, there is also a technical aspect concerning the type of tools and mechanisms and how to engage with the needs of farmers and the challenges that agriculture faces—as an economic activity, a social practice, and a type of land use (farmland).

Despite social and cultural struggles, agriculture and food production are essential activities occurring in most peri-urban areas across Europe. As repeatedly mentioned, agriculture proves to be a cultural and economic asset for local communities and an environmental one when attention is given to the detrimental effects of certain types of industrial and intensive production. The park as a mechanism remains a valid resource. However, it needs to be critically addressed again to move beyond simplistic forms of farmland protection and to demonstrate its capacity to dynamize farms' activities and agrarian communities by establishing more metabolic relationships across scales. Ecological features are a driver for integrating these parks within a more coherent context, many of which are crucial for agricultural activities, such as in the case of the Baix Llobregat and Emscher Landscape parks. Therefore, the design process needs to acknowledge and incorporate these features. By "design process," it refers to planning and design as a collective activity within a broad field of action, capable of integrating various actors, interests, and technical competencies into a vision that could later guide implementation, as well as the management, care, and maintenance of these territories. The contributions made by the Territorialists in Tuscany are key references for this definition.

For this reason, this chapter intentionally revisits some of these arguments by directly studying selected agrarian territories to understand the type of planning, management, and governance processes needed to address the twofold needs of these territories: the need to protect them from various kinds of pressures and the need to recognize their needs as economic activities and small enterprises. A third element, and perhaps one that has more recently emerged as a priority, concerns identifying the appropriate mechanisms to guide these territories and communities toward a socio-ecological transition through their practices and forms of consumption.

The following chapters build on this historically oriented research, which provides a solid foundation for studying and analyzing these territories in the present context and through the thematic lenses proposed in this book.

References

Balducci, A., & Piazza, M. (1981). *Dal Parco Sud al cemento armato. Politica urbanistica e strategie immobiliari nell'area milanese*. ACLI.

Brandt, J., & Vejre, H. (2004). *Multifunctional landscapes—Motives, concepts and perceptions*. In J. Brandt & H. Vejre (Eds.), Multifunctional Landscapes (pp. 3–32). WIT Press.

Calori, A., & Magarini, A. (2015). *The role of informal food networks: Notes from the experience of Milan toward food policy councils*. FAO Food and Nutrition Paper.

Donadieu, P. (2013). *Campagne urbane. Una nuova proposta di paesaggio della città* (M. Mininni, Ed.; Vol. 102). Donzelli Editore.

Esteves, M. J. (2022). El Parque Agrario en la planificación del crecimiento urbano en tierras secas: El caso de Guaymallén, Mendoza (Argentina). *Revista de Urbanismo, 47*. https://doi.org/10.5354/0717-5051.2022.66848

Fanfani, D. (2006). Il governo del territorio e del paesaggio rurale nello spazio "terzo" periurbano. Il parco agricolo come strumento di politiche e di progetto. *Ri-Vista. Research for* Landscape Architecture, *6*(2), Article 2. https://doi.org/10.13128/RV-17443.

Fanfani, D. (2009). Lo scenario progettuale per il parco agricolo di Prato: Un laboratorio in progress di politiche. In *Pianificare tra città e campagna: Scenari, attori e progetti di nuova ruralità per il territorio di Prato. - (Strumenti per la didattica e la ricerca ; 88)* (pp. 1000–1021). Firenze University Press.

Fanfani, D. (2019). Agricultural Park in Europe as tool for Agri-urban policies and design: A critical overview. In E. Gottero (Ed.), *Agrourbanism: Tools for governance and planning of agrarian landscape* (pp. 149–169). Springer International Publishing. https://doi.org/10.1007/978-3-319-95576-6_10

Fanfani, D., & Matarán Ruiz, A. (Eds.). (2020). *Bioregional planning and design: Volume I: Perspectives on a transitional century*. Springer International Publishing. https://doi.org/10.1007/978-3-030-45870-6

Ferraresi, G., & Rossi, A. (Eds.). (1993). *Il parco come cura e coltura del territorio. Un percorso di ricerca sull'ipotesi del parco agricolo*. Grafo edizioni.

Gallent, N. (2006). The rural–urban fringe: A new priority for planning policy? *Planning Practice & Research, 21*(3), 383–393. https://doi.org/10.1080/02697450601090872

Gkartzios, M., Gallent, N., & Scott, M. (2022). *Rural places and planning. Stories from the global countryside*. Bristol University Press.

Gruehn, D. (2017). Regional planning and projects in the Ruhr region (Germany). In M. Yokohari, A. Murakami, Y. Hara, & K. Tsuchiya (Eds.), *Sustainable landscape planning in selected urban regions* (pp. 215–225). https://doi.org/10.1007/978-4-431-56445-4_18

Jarrige, F., & Perrin, C. (2017). L'agriparc: Une innovation pour l'agriculture des territoires urbains? *Revue d'Économie Régionale & Urbaine, Juin*(3), 537–562. https://doi.org/10.3917/reru.173.0537.

Keil, R., & Macdonald, S. (2016). Rethinking urban political ecology from the outside in: Greenbelts and boundaries in the post-suburban city. *Local Environment, 21*(12), 1516–1533. https://doi.org/10.1080/13549839.2016.1145642

Kipar, A. (1993). *L'Emscher-Park nel bacino della Ruhr*. Il Pomerio.

Lemes de Oliveira, F. (2019). Green wedges: The resilience of a planning idea. In F. Lemes de Oliveira & I. Mell (Eds.), *Planning cities with nature: Theories, strategies and methods* (pp. 17–27). Springer International Publishing. https://doi.org/10.1007/978-3-030-01866-5_2

Lohrberg, F., Licka, L., Scazzosi, L., & Timpe, A. (Eds.). (2015). *Urban Agriculture Europe*. Jovis.

Magnaghi, A. (2000). *Il progetto locale*. Bollati Boringhieri.

Magnaghi, A. (2011, February 21). *Territorialists' Society Manifesto* (M.-A. Gillis, Trans.) Società dei Territorialisti SdT. https://www.societadeiterritorialisti.it/wpcontent/ uploads/2013/05/110221b_draft.of.the.territorialists.society.manifesto.pdf

Magnaghi, A., & Fanfani, D. (Eds.). (2010). *Patto città campagna: Un progetto di bioregione urbana per la Toscana centrale*. Alinea. http://hdl.handle.net/2158/608730

Mantovani, G. V., Peralta Flores, M. C., & Soijet, M. G. C. (2020, June). La figura del parque agrario en Argentina: Un recurso para afrontar las problemáticas de la urbanización metropolitana. In *XII Seminario Internacional de Investigación en Urbanismo, São Paulo-Lisboa, 2020*. Seminario Internacional de Investigación en Urbanismo. https://doi.org/10.5821/siiu.9923

Matros, J., & Lohrberg, F. (2016). Eine Plantage als Park? Begleitforschung zur Beerntung einer Kurzumtriebsplantage im Waldlabor Köln. *Naturschutz und Landschaftsplanung*, *48*(1), 15–21.

Mininni, M. (2013). *Approsimazioni alla città. Urbano, rurale, ecologia*. Donzelli Editore.

Olmo, R., & Yacamán Ochoa, C. (2015). *Gobernanza para una agricultura viva en un paisaje periurbano de calidad. Estudios de caso en la huerta metropolitana de Madrid* (pp. 265–274).

Paül, V. (2015). Los parques agrarios en Cataluña. Breve análisis de la contribución de Josep Montasell a su concepción, desarrollo e implantación. In *El Parque Agrario. Una figura de transición hacia nuevos modelos de gobernanza territorial y alimentaria*. (pp. 113–142). Heliconia S. Coop. Mad.

Poli, D. (2015). La nuova categoria di spazio pubblico territoriale nella bioregione urbana fra parco agricolo multifunzionale e contratto di fiume. *Radici, Condizioni, Prospettive*, 8.

Poli, D. (2023). Il paesaggio: Patrimonio storico fondante della pianificazione territoriale. *Contesti. Città, Territori, Progetti*, 240–253. https://doi.org/10.36253/contest-14882.

Rettig, S. (1976). An investigation into the problems of urban fringe agriculture in a greenbelt situation. *Planning Outlook*, *19*(1–2), 50–74. https://doi.org/10.1080/00320717608711519

Rocamonde, J., & Sabaté Bel, J. (2019). Paisajes Cooperativos. Una comparativa de dos paisajes industriales vinculados al río Llobregat. Ciudades, (22), Article 22. https://doi.org/10.24197/ciudades.22.2019.151-172.

Sabaté Bel, J. (2000). El Parc Agrari del Baix Llobregat. *Àrea Revista de Debats Territorials*, 251–282.

Sabaté Bel, J. (2015). Reflexiones en torno al proyecto urbanístico de un Parque Agrario. In *El Parque Agrario: Una figura de transición hacia nuevos modelos de gobernanza territorial y alimentaria* (pp. 93–112). Heliconia. http://hdl.handle.net/2117/83030

Salvador, A. J. (2024a). Defining a Peri-urban agrarian space as a socio-spatial and environmental unit. Lessons from Barcelona's local food system. In F. Calabrò, L. Madureira, F. C. Morabito, & M. J. Piñeira Mantiñán (Eds.), *Networks, Markets & People* (pp. 72–83). Springer Nature. https://doi.org/10.1007/978-3-031-74704-5_8

Salvador, A. J. (2024b). Unpacking Parco Agricolo Sud through a socio-spatial production perspective. *TERRITORIO*, (2023/104). https://doi.org/10.3280/TR2023-104007.

Salvador, A. J. (2025). Re-programming agrarian urbanism in South Milan. The peri-urban question. *Landscape Research, 50*(8), 1303–1318. https://doi.org/10.1080/01426397.2025.2526497

Timpe, A. (2017). *Produktive Parks entwerfen. Geschichte und aktuelle Praxis biologischer Produktion in europäischen Parks.* [Doctoral Thesis, RWTH Aachen]. urn:nbn:de:hbz:82-rwth-2017-005402.

Timpe, A. (2019). *Designing Productive Parks: From Romanticizing Agriculture towards Green Infrastructure and Productive Public Space: The Ruhr Metropolis as a Laboratory for Productive Parks* [Print]. LIT Verlag.

Timpe, A. (2020). Cultivating the Cologne Green Belt: The belvedere Agricultural Park. In L. Scazzosi & P. Branduini (Eds.), *AgriCultura: Urban agriculture and the heritage potential of agrarian landscape* (pp. 205–224). Springer International Publishing. https://doi.org/10.1007/978-3-030-49012-6_13

Van der Ploeg, J. D., Laurent, C., Blondeau, F., & Bonnafous, P. (2009). Farm diversity, classification schemes and multifunctionality. *Journal of Environmental Management, Multifunctional Agriculture - From Farm Diagnosis to Farm Design and Institutional Innovation, 90*, S124–S131. https://doi.org/10.1016/j.jenvman.2008.11.022

Vejre, H., Abildtrup, J., Andersen, E., Andersen, P. S., Brandt, J., Busck, A., Dalgaard, T., Hasler, B., Huusom, H., Kristensen, L. S., Kristensen, S. P., & Præstholm, S. (2007). Multifunctional agriculture and multifunctional landscapes—Land use as an interface. In Ü. Mander, H. Wiggering, & K. Helming (Eds.), *Multifunctional land use: Meeting future demands for landscape goods and services* (pp. 93–104). Springer. https://doi.org/10.1007/978-3-540-36763-5_6

Verzone, C., & Woods, C. (2021). Food Urbanism: Typologies, Strategies, Case Studies. In *Food Urbanism*. Birkhäuser. https://doi.org/10.1515/9783035615678.

Vescovi, F. (2012). *Proposte per il Parco Agricolo Sud Milano. Criticità e risorse dell'agricoltura periurbana*. Ronca Editore.

Whitehand, J. W. R., & Morton, N. J. (2006). The fringe-belt phenomenon and socioeconomic change. *Urban Studies, 43*(11), 2047–2066. https://doi.org/10.1080/00420980600945138

Yacamán Ochoa, C., & Zazo Moratalla, A. (Eds.). (2015). *El Parque Agrario. Una figura de transición hacia nuevos modelos de gobernanza territorial y alimentaria.* Heliconia S. Coop. Mad.

Zasada, I. (2011). Multifunctional peri-urban agriculture—A review of societal demands and the provision of goods and services by farming. *Land Use Policy, 28*(4), 639–648. https://doi.org/10.1016/j.landusepol.2011.01.008

Zazo Moratalla, A., & Paül, V. (2022). What is an Agricultural Park? Observations from the Spanish experience | Elsevier enhanced reader. *Land Use Policy, 112.* https://doi.org/10.1016/j.landusepol.2021.105584

Zepp, H. (2018). Regional green belts in the Ruhr region. A planning concept revisited in view of ecosystem services. *Erdkunde, 72.* https://doi.org/10.3112/erdkunde.2018.01.01

CHAPTER 4

A Differentiated Agricultural Production

Introduction

Before delving into the various themes proposed by this book, it is important to describe the agricultural activity in each of the three contexts. This is crucial for several reasons, as agriculture encompasses economic, cultural, and environmental dimensions that are both significant and differentiated across the cases. Not only do the challenges faced by each territory vary, but so do the demographics of farmers, conditions for accessing farmland, availability of natural resources, and other contextual factors. While some of these topics are addressed in greater depth in individual chapters, the aim of this chapter is to provide an overview of what is produced in each case, with particular attention to the limitations that constrain diversified production and its capacity to meet consumer needs while supporting a transition toward more sustainable farming practices and local food systems.

This chapter primarily draws on data gathered from national and European sources. It is important to emphasize that its purpose is not to establish a direct comparison between cases, but rather to highlight the complexity of each context and the plurality of agricultural production in European peri-urban areas. A comparative analysis is neither deemed necessary nor fully feasible, given the limitations and inconsistencies among available datasets. For the purposes of this analysis, it is therefore more relevant to identify the specific characteristics and challenges of each

© The Author(s), under exclusive license to Springer Nature Switzerland AG 2026
A. J. Salvador, *Metropolitan Food Systems*, Sustainable Urban Futures, https://doi.org/10.1007/978-3-032-26016-1_4

context. One of the key challenges for planners, policymakers, and anyone attempting to integrate food and agrarian issues into broader debates on cities, the climate crisis, and socio-ecological transitions is understanding the scale of production and consumption, and how these scales intersect with administrative and governance boundaries.

Anticipating some of the findings, the Barcelona case reveals a highly diversified agricultural production, particularly when comparing coastal production areas with those located inland. It demonstrates how farm size, land tenure, and tendencies toward specific types of production vary and respond to the market conditions associated with different crops. The Ruhr Metropolis case shows that agriculture has not traditionally been oriented toward food consumption but rather toward the production of commodities, although this orientation appears to be evolving. This has implications for farm size, as many holdings are significantly larger and geared toward industrial-scale production. Finally, the Milan case illustrates a territory that, while diversified at a broader regional scale—especially in areas characterized by distinct topographical and natural features—exhibits highly specialized and spatially concentrated production systems, often monocultural in nature. For example, production areas closer to Milan are strongly specialized in rice cultivation and dairy production. This specialization highlights long-standing agricultural activities that have evolved and adapted to globalized markets, reinforcing an industrialized and specialized production model.

Territorial Differentiation and Agricultural Diversity in the Barcelona Metropolitan Area

Barcelona's farming style along its coast is known as "*huerta*," referring to the historic areas of vegetable and fruit production intended to supply nearby cities along the Spanish Mediterranean coastline. These areas often maintain a direct connection with rivers linking the mountains to the sea. As Ostrom (1990, pp. 69–81) argues, the historic *huerta* irrigation institutions along the Spanish Mediterranean coast are an enduring example of common-good governance—particularly in water management—fostering a strong sense of responsibility and stability in these contexts. In the specific case of the Baix Llobregat Delta, the irrigation systems that enabled the transition toward horticultural production were promoted and financed by a group of farmers in the early nineteenth century. This process dramatically transformed agricultural activity, turning the delta into one of Europe's main vegetable production areas (Rocamonde & Sabaté Bel, 2019).

According to 2020 data from the Instituto Nacional de Estadística (INE), the Municipality of Barcelona has over 1.6 million inhabitants. The broader Province of Barcelona, which includes 311 municipalities across coastal and inland areas, is home to more than 5.6 million people. As noted by the Diputació de Barcelona (2015), the metropolitan de facto urban area extends beyond administrative boundaries, forming a continuous urbanized corridor along the coast and inland valleys that connects to nearby cities such as Tarragona, Girona, and Manresa. This Functional Urban Area (FUA) encompasses roughly 4.8 million people, representing the core of Catalonia's economic, cultural, and transport networks. The entire Catalonia region has a population of approximately 7.8 million, highlighting the dense interrelation between urban and peri-urban territories where agricultural production remains active.

As anticipated earlier, one of the most interesting features of agricultural production in Barcelona's metropolitan area is the way production patterns vary by county (*comarca*) and in relation to proximity to the sea (see Fig. 4.1). For this analysis, four counties have been selected to highlight these differences: Baix Llobregat and El Maresme, which share a strong horticultural orientation and are located along the lower stretches of rivers as they meet the sea; Penedès, known for wine production, and Anoia, which is specialized in cereals and grains.

As in other European metropolitan contexts, understanding agricultural production in the Barcelona Metropolitan Area requires considering both the diversity of crops and the territorial conditions in which they occur. Agriculture here is strongly shaped by topography, proximity to the sea, and historical specialization, producing a varied landscape of intensive horticulture, vineyards, and cereal production. Table 4.1 shows that the total Utilized Agricultural Area (UAA) in the four selected counties—Baix Llobregat, Maresme, Penedès, and Anoia—ranges from 3403 ha in Maresme to 31,012 ha in Anoia, with average farm sizes between 5 ha in Maresme and 28 ha in Anoia. The region reflects both small-scale, high-value production near the coast and larger, specialized farms inland.

The coastal counties of Baix Llobregat and Maresme are both primarily oriented towards horticulture, yet their production methods differ significantly due to territorial and regulatory conditions. In Baix Llobregat, the presence of the agricultural park shapes agricultural practices by restricting

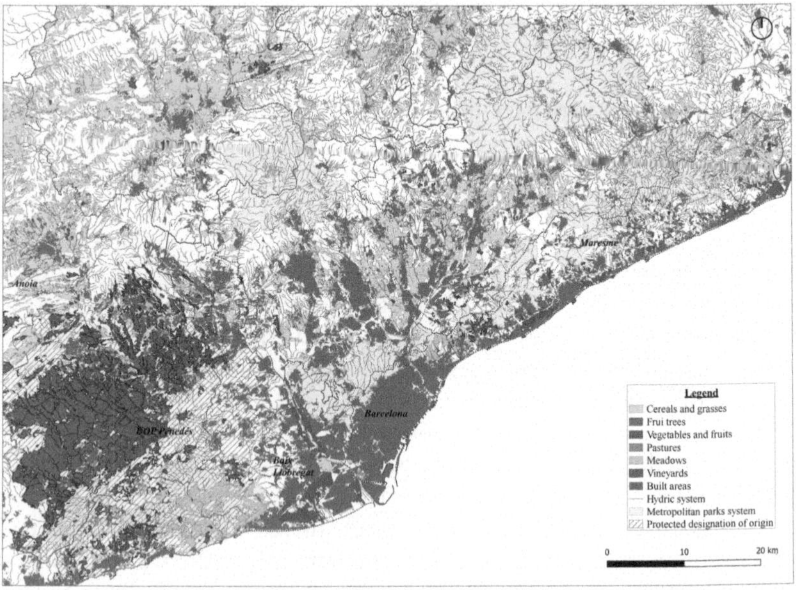

Fig. 4.1 Agricultural production in the Barcelona metropolitan area. The counties discussed in the text are evidenced. (Source: Elaborated by the author with data from GenCat)

greenhouse cultivation to protect the landscape, influencing production cycles, biodiversity, and business models. The county accounts for 4018 ha of UAA, dominated by fruits and vegetables, managed largely on small farms averaging 7 ha, which reflects an intensive production system focused on local markets and high-value crops. In contrast, Maresme, with 3403 ha of UAA, emphasizes greenhouse vegetable production—122 ha of which, roughly 20% of vegetable farmland, are under greenhouses—illustrating both the intensity of cultivation and the reasons behind local farmers' resistance to establishing a new agricultural park, as noted by staff at the Barcelona Provincial Council (*Diputació de Barcelona*). Overall, greenhouses occupy a significant share of production in Maresme, including vegetables and ornamental plants, while ecological management covers a smaller portion (8–9% of total UAA), highlighting a highly specialized and intensive coastal horticultural landscape.

Table 4.1 Agricultural production in the Barcelona metropolitan area (Unit of measure: hectares)

	Usable agricultural area (UAA)	*Average*	*Main*	*Share of economic accounts for agriculture (EAA)*	*UAA with ecological conduction*	*% of total UAA*
Baix Llobregat	4.018 ha	7 ha	Fruits and vegetables	0.12%	368 ha	9.17%
Maresme	3.403 ha	5 ha	Vegetables (greenhouses)	0.07%	277 ha	8.15%
Penedès	25.055 ha	13 ha	Vineyards	0.21%	10.097 ha	40.31%
Anoia	31.012 ha	28 ha	Cereals and grain	0.21%	2.779 ha	8.96%
Catalunya	**1.092.215 ha**	**22 ha**		**1151%**	**161.286 ha**	**1477%**
Spain	**23.913.682 ha**	**26 ha**		**100%**	**1.871.529 ha**	**783%**

Source: Elaborated by the author with data from the 2020 Agricultural Census by the Instituto Nacional de Estadística

Not all Baix Llobregat county data corresponds to the agricultural park itself. Within the park, there are 318 farm holdings over 1924 hectares of UAA, representing 55% of the park's total area (3498 hectares). According to the Institut Agrícola Català de Sant Isidre,[1] the number of holdings has declined sharply since the park's establishment in 1998: from 2102 in 1982 to 621 in 1999, and 318 in 2020—a 49% reduction. Vegetable production dominates the park, with 1075 hectares, of which only 53 hectares are under roofed structures. Approximately 1200 people work in the park, producing 35,500 tons of fruits and vegetables annually.[2] Renowned products include artichokes from El Prat, asparagus from Gavà and Viladecans, chards, leeks, and Catalan scallions (*calçots*) from Baix Llobregat, and cherries from Santa Coloma de Cervelló, Livestock farming is limited but present, with 23 holdings raising poultry or sheep. The

[1] The Institute Agrícola Catalá de Sant Isidre is the oldest farmers' association in Spain, the report can be accessed at: https://institutagricola.org/es/el-parque-agrario-un-ejemplo-de-nefasta-gestion-del-potencial-agroalimentario-del-delta-del-llobregat/. Last visited on 25/02/2025.

[2] Data provided by the Baix Llobregat agricultural park management entity during a meeting held in June 2024.

Pota Blava chicken, a domestic breed from El Prat de Llobregat, holds PGI protection.

In contrast, the inland counties of Penedès and Anoia show a different model of agricultural specialization. Penedès is dominated by vineyards, covering 19,380 ha, representing over 40% of the county's UAA, with a significant portion under ecological management (10,097 ha, 40% of total UAA), reflecting a growing trend in organic viticulture. Anoia, by contrast, is the main cereal-producing area, with 20,793 ha devoted to cereals and grains, representing the majority of its UAA, and with much larger average farm sizes of 28 ha. These differences illustrate how topography, soil, and historical land use shape both the scale and type of production, as well as the potential for diversified versus specialized agriculture.

Tables 4.2 and 4.3 further highlight the diversity of production within each county. Baix Llobregat combines cereals (408 ha), vegetables (891 ha), stone fruits (395 ha), grapes (258 ha), olives (324 ha), and permanent grasslands (923 ha), illustrating a mixed production landscape. Maresme shows a more horticultural focus, with 639 ha of vegetables and 417 ha of plants harvested green, alongside 275 ha of grassland. Penedès is almost exclusively viticultural, while Anoia is strongly cereal based, with minimal vegetable production. Greenhouse cultivation is concentrated near the coast, particularly in Maresme, whereas inland counties rely on extensive open-field production.

This territorial differentiation also intersects with economic and ecological aspects. Despite the small share of total economic accounts for agriculture (0.07–0.21% at the county level), ecological management has been implemented over a considerable area, especially in Penedès, reinforcing the role of sustainable practices within the wine industry. The contrast between small, intensive coastal farms and larger, specialized inland operations reflects not only production strategies but also the social and landscape implications of agriculture in metropolitan Barcelona. Intensive horticulture produces highly managed, heterogeneous landscapes with significant visual and ecological implications, while large cereal or vineyard areas create more homogeneous rural landscapes but are essential for regional production and economic specialization.

Barcelona's metropolitan agricultural sector is distinctive within Catalonia and Spain in terms of both farm types and production diversity. The territory is differentiated into areas with specific characteristics, vulnerabilities, and historical, ecological, and socioeconomic contexts, offering valuable insights for planning and governance as well as providing

Table 4.2 Type of agricultural production in the Barcelona metropolitan area (Unit of measure: hectares)

	Cereals	*Vegetables*	*Stone fruits (temperate climate)*	*Grapes*	*Olives*	*Plants harvested green*	*Permanent grassland*
Baix Llobregat	408 ha	891 ha	395 ha	258 ha	324 ha	200 ha	923 ha
Maresme	906 ha	639 ha	27 ha	242 ha	29 ha	417 ha	275 ha
Penedès	1.612 ha	232 ha	345 ha	19.380 ha	824 ha	251 ha	437 ha
Anoia	20.793 ha	60 ha	118 ha	3.021 ha	523 ha	351 ha	710 ha
Barcelona Province	**72.097 ha**	**2.192 ha**	**930 ha**	**23.503 ha**	**2.570 ha**	**21.143 ha**	**45.533 ha**

Source: Elaborated by the author with data from the 2020 Agricultural Census by the Instituto Nacional de Estadística

Table 4.3 Type of agricultural production in the Barcelona metropolitan area (Unit of measure: hectares)

	Orchards for self-consumption	*Vegetables (greenhouse)*	*Flowers and plants (greenhouse)*	*Total UAA in greenhouses*	*Total UAA uncultivated land*	*Forest area*
Baix Llobregat	2 ha			50 ha	95 ha	819 ha
Maresme	3 ha	122 ha	77 ha	212 ha	37 ha	1.804 ha
Penedès	6 ha	118 ha	0 ha	122 ha	44 ha	9.652 ha
Anoia	5 ha	4 ha	0 ha	4 ha	5 ha	10.255 ha
Barcelona Province	**42 ha**	**293 ha**	**86 ha**	**411 ha**	**252 ha**	**81.598 ha**

Source: Elaborated by the author with data from the 2020 Agricultural Census by the Instituto Nacional de Estadística

insights of peri-urban agriculture within broader metropolitan planning strategies.

Agricultural Transition and Spatial Differentiation in the Ruhr Metropolis

As in the previous case, profiling the Ruhr Metropolis through its agricultural activity, farmers, and farm business models is essential for understanding how this production fits within this project's aim to analyze local food systems in European metropolitan countryside areas. The analysis draws on data from the Eurostat database and publications by the Regionalverband Ruhr—RVR (Handelsblatt Research Institute, 2021) and the North Rhine-Westphalia Chamber of Agriculture (Landwirtschaftskammer Nordrhein-Westfalen, 2012, 2019). The 2012 document is particularly important, as it forms a core part of the results produced by the KuLaRuhr project[3] and has played a significant role in policymaking, governance, and political lobbying regarding the role and importance of agriculture in the Ruhr Metropolis area. Eurostat data are organized by NUTS 2 regions, which do not correspond exactly to the area defined as the RVR, covering a larger territory.[4] Nevertheless, this data provides valuable insights into the types of production in each region. More localized data, tailored to the area under analysis, are provided by the NRW Chamber of Agriculture and the RVR.

According to 2020 data from the Landesbetrieb Information und Technik Nordrhein-Westfalen (IT.NRW), the central core of the Ruhr Metropolis includes 11 *kreisfreie Städte*, with Dortmund, Essen, and Duisburg being the largest, totaling over 5.1 million residents in a total area of 4.439 km^2 (443.900 hectares), of which 39% are urbanized areas, and 61% corresponds to natural areas, including water bodies. This gives a population density of 1.148 people per sq.km, forming one of the largest

[3] The project's official name in German is *KuLaRuhr — Nachhaltige urbane Kulturlandschaft in der Metropole Ruhr*. It was a joint initiative funded by the German Federal Ministry of Education and Research (*Bundesministerium für Bildung und Forschung*, BMBF). Although the project concluded in 2014, a repository of project documentation and outputs remains available online: http://www.kularuhr.de/index.php/projekt.html. Last accessed on 08/03/2025.

[4] The analysis focuses on the NUTS 2 region of North Rhine–Westphalia (DEA), specifically the NUTS 3 units of Düsseldorf (DEA 1), Münster (DEA 3), and Arnsberg (DEA 5), encompassing their major cities and surrounding districts.

urban areas in Europe. The Ruhr Metropolis is a polycentric urban agglomeration that can be defined by different dimensions: a legal one which describes this as one of Germany's metropolitan regions, an economic one primarily determined by its historical mining industry activities, a planning one that links the previous both through the role of the Regional Association Ruhr (*Regionalverband Ruhr—RVR*), by an administrative dimension represented by 53 towns and cities, by a morphological one defined by a continuous and heterogeneous urban fabric, and by a geographical one of the Ruhr Valley defined by the blue infrastructure of the Ruhr, Lippe, and Emscher rivers system and a human-made canal infrastructure that ties it with the other dimensions (Fig. 4.2). This urbanized area is a continuous mesh of cities, industrial zones, and transport infrastructure, yet it remains interspersed with agricultural land, forests, and watercourses.

According to the 2019 NRW Chamber of Agriculture report (Landwirtschaftskammer Nordrhein-Westfalen, 2019), the total Utilized

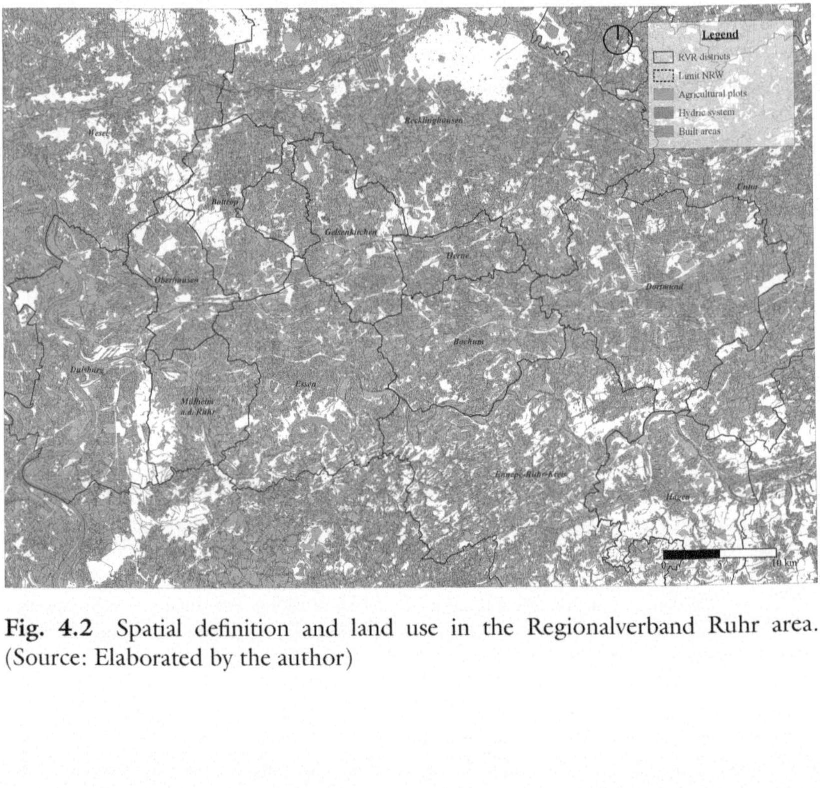

Fig. 4.2 Spatial definition and land use in the Regionalverband Ruhr area. (Source: Elaborated by the author)

Agricultural Area (UAA) in 2019 was 165,002 ha. The report also notes that the RVR region loses approximately 1246 ha of agricultural land each year, and agriculture is no longer the primary land use as it was in 2012. In 2012, agriculture accounted for 39.2% of the RVR's jurisdictional area, slightly above urbanized areas (38.5%), followed by forests (18.2%) and water infrastructure (3.2%) (Landwirtschaftskammer Nordrhein-Westfalen, 2012). By 2017, agriculture had fallen to the second-largest land use (37.2%), behind urbanized areas and mobility infrastructure (39.1%). Dortmund and Bochum, in the central core of the Ruhr Metropolis, experienced the greatest farmland loss between 2010 and 2017, followed by Gelsenkirchen, Herne, and Duisburg.

According to the Regionalverband Ruhr (2023), compensation measures for new infrastructure or developments often result in additional farmland loss, as some farmland is converted into green areas or future woodlands. This has created a *double loss* of farmland, primarily because farmland is cheaper and easier to use for compensation measures. Outer municipalities frequently bear the burden of this loss, as compensation activities often take place in locations different from the original project site. The 2019 NRW Chamber of Agriculture report confirms that these compensation measures, along with new large-scale retail centers and gravel extraction, are major reasons for the high farmland loss in the outer municipality of Wesel, which lost over 1440 ha of farmland between 2010 and 2017. This aspect will be discussed in greater detail in subsequent chapters, as it constitutes one of the strategies adopted by large institutional farmland owners.

Table 4.4 presents 2015 land-use data for the 11 urban municipalities[5] or independent cities (*kreisfreie Städte*) of the Ruhr Metropolis, mostly in the central core (Fig. 4.2). Hamm has the largest agricultural area at 12,185 ha, representing 53.8% of its total area. Within the central core, Dortmund has 6714 ha, Essen 3328 ha, and Bottrop 2922 ha. The smallest agricultural surfaces are in Herne (481 ha), Oberhausen (680 ha), and Gelsenkirchen (1467 ha). Bottrop (29.0%), Dortmund (23.9%), and Mülheim an der Ruhr (22.9%) are the only municipalities with more than 20% of their area dedicated to agriculture. The average across the 11 municipalities is 23.9%.

[5] These municipalities are Bochum, Bottrop, Dortmund, Duisburg, Essen, Gelsenkirchen, Hagen, Hamm, Herne, Mülheim a. d. Ruhr, Oberhausen.

Table 4.4 Land use distribution in the Ruhr Metropolis (Unit of measure: hectares)

	Total surface	*Agricultural land*	*Agr. land/ total surface %*	*Total UAA 2011*[a]	*Forests*	*Water surface*	*Recreational and green areas*	*Mining land*	*Built area*	*Mobility infrastructure*
Urban municipalities of Ruhr Metropolis	**168,173**	**40,256**	23.9	33,968	**25,772**	**4843**	**12,009**	**279**	**55,573**	**23,280**
Duisburg	23,280	4422	*19.0*	3112	2047	2023	1800	9	8368	3657
Essen	21,034	3328	*15.8*	3471	2858	569	2352	0	8035	2945
Mülheim an der Ruhr	9128	2091	*22.9*	1455	1742	200	666	3	2903	1267
Oberhausen	7709	680	*8.8*	444	1055	73	829		3324	1364
Bottrop	10,061	2922	*29.0*	3448	2328	226	563	115	2316	1075
Gelsenkirchen	10,494	1467	*14.0*	1008	944	314	1081		4438	1630
Bochum	14,566	2879	*19.8*	1920	1228	175	1426	1	6161	2188
Dortmund	28,071	6714	*23.9*	5318	3996	243	1701	9	10,088	4420
Hagen	16,045	3087	*19.2*	1912	6733	361	438	138	3327	1579
Hamm	22,643	12,185	*53.8*	11303	2294	555	602	4	4387	2086
Herne	5142	481	*9.4*	577	547	104	551	0	2226	1069
North Rhine-Westphalia	**3,411,252**	**1,646,384**	48.3		**887,765**	**66,155**	**70,827**	**19,633**	**437,704**	**245,476**
Germany	**35,740,911**	**18,433,248**	51.6		**10,951,461**	**855,213**	**445,484**	**155,871**	**2,507,664**	**1,810,805**

Source: Elaborated by the author with NRW Chamber of Agriculture data

Source: Federal and State Statistical Offices of Germany

[a]Landwirtschaftskammer Nordrhein-Westfalen, 2012

According to the 2019 NRW Chamber of Agriculture report (using 2015 data), cereals are the main crop in the Ruhr Metropolis, covering over 55,000 ha. Maize follows with just over 20,000 ha, rapeseed with 5000 ha, potatoes with approximately 2000 ha, and sugar beet with around 1000 ha. While grain production had been increasing in 2016, industrial crops such as rapeseed, sunflower, and sugar beet have gained importance in recent years due to the Ukraine-Russia war. Regional data indicates that overall oilseed production in Germany significantly increased in 2022 and 2023 compared to 2019–2021, both in area and output, though it has not yet reached the peaks of 2013–2015.

Grasslands are also an important agricultural land use, especially in outer municipalities. For example, in Herdecke and Breckerfeld in the southeast, nearly 80% of agricultural land is grassland, while other municipalities exceed 50%. In the central core, the distribution varies, eastern municipalities like Herne and Dortmund are predominantly arable, while western municipalities such as Essen and Oberhausen are mainly forested with small agricultural patches, mostly grasslands. Considering the entire state of North Rhine-Westphalia, Table 4.5 shows that in 2010 and 2020, 71% of the UAA was arable land (ARA) and 27% was permanent grassland. Among arable land, cereals for grain production dominated (61.28% in 2010, 54.41% in 2020), followed by plants harvested green—primarily silage and green maize (19.65% in 2010, 25.51% in 2020). Root crops were the third most important crop (8.26% in 2010, 8.38% in 2020). Industrial crops were fourth and declined until 2020, consistent with trends from 2015–2021, with production rebounding during the war.

By 2020, the production of vegetables, strawberries, other garden crops, and legumes had increased across North Rhine-Westphalia. Data from the NRW Chamber of Agriculture reveals an increase of 16.9% in vegetables, strawberries, and other garden crops, with top fruits and berries adding roughly 4000 ha more. Field beans and peas accounted for roughly 90% of legume output, while strawberries were particularly significant both economically and socially. For example, Hof Mertin farm in Dortmund combines farm sales, city distribution, pick-your-own activities, and commercial production for supermarkets and exports. According to the 2019 NRW Chamber of Agriculture report, approximately 418 hectares of strawberries are cultivated in the Ruhr Metropolis.

While cereals and industrial crops require extensive land, vegetables and strawberries occupy smaller areas, highlighting different production strategies. From a landscape perspective, extensive monocultures create

Table 4.5 Breakdown of the total farm area (TA) in North Rhine-Westphalia in 2020 (Unit of measure: holdings and hectares)

		Farm holdings	*Area (ha)*	*% of Total UAA*	*% of Total ARA*	*Var. farm holdings 2010–2020*	*Var. area 2010–2020*
2020	**Utilized agricultural area (UAA)**	**31,941**	**1,473,157**			**−9.97%**	**0.69%**
	Arable land (ARA)	**25,479**	**1,047,873**	**71.13%**		**−13.35%**	**−0.42%**
	Cereals for the production of grain	20,769	570,124		54.41%		
	Plants harvested green	17,507	267,291		25.51%		
	Root crops	5919	87,849		8.38%		
	Legumes for grain production	1794	18,215		1.74%		
	Industrial crops	4149	44,527		4.25%		
	Vegetables, strawberries and other garden plants	2712	31,531		3.01%		
	Seed and seedling products, grasses, root-crops etc.	159	1412		0.13%		
	Other crops on arable land	712	1770		0.17%		
	Set aside areas, fallow land	10,752	25,154		2.40%		
	Permanent crops (PECR)	**1243**	**12,173**	**0.83%**		**−30.36%**	**−8.17%**
	Permanent grassland	**26,964**	**411,497**	**27.93%**		**−2.05%**	**3.71%**
	Kitchen gardens (gardening land)	**3188**	**1615**	**0.11%**		**−3.98%**	**126.51%**
	Other areas	**24,167**	**156,278**				
	Agr. area excl	1504	5037				
	Wooded areas (WA) and short-rotation coppices (SRCAA)	13,391	130,057			**−10.57%**	**−16.77%**
	Land occupied by buildings, farmyards, other areas	20,949	21,184			**−3.56%**	**2.34%**
	Special holdings areas (SAA)—mushrooms	**12**	**6**				
	Total farm area (FA)	**33,611**	**1,629,435**			**−5.98%**	**−0.71%**

Source: Statistisches Bundesamt (Destatis)

Source: Elaborated by the author with Destatis data

homogeneous environments, whereas diversified production supports ecological and social benefits. Wheat, silage/green maize, and barley are the most widespread crops, cultivated by over half of all farms, while some crops, such as industrial plants or sugar beet, are not intended for human consumption.

Permanent grasslands account for 27.93% of UAA, yet 84.42% of farms maintain meadows, pastures, or low-yield grasslands for grazing. The Ruhr Metropolis hosts 181,680 livestock units, with animal husbandry—particularly horse breeding in Düsseldorf and Arnsberg—playing a key economic and social role. Horticulture represents 12% of total agricultural production in the Ruhr Metropolis, double the state average of 6%, and revenue per hectare is higher than the German average due to intensive vegetable production (Handelsblatt Research Institute, 2021).

Finally, data reveal that although total UAA in North Rhine-Westphalia increased by 0.69%, the number of holdings dropped by 9.97%, affecting average farm size. Land for permanent crops decreased by 8.17%, while wooded areas and short-rotation coppices fell by 16.77%. Holdings engaged in permanent crops declined by 30.36%, primarily encompassing top fruits, berries, tree nurseries, Christmas tree plantations, and vineyards

The agricultural landscape of the Ruhr Metropolis is highly differentiated, reflecting both historical patterns and evolving economic pressures. Eastern municipalities such as Dortmund and Herne remain predominantly arable, with cereals, maize, and silage crops dominating, while western municipalities like Essen and Oberhausen are largely forested, with small patches of grassland and specialized horticulture. Outer municipalities show a predominance of grasslands, supporting livestock and dairy farming, and highlighting the importance of animal husbandry as a social and economic activity. Over time, the Ruhr's production has shifted, cereals and industrial crops still occupy extensive areas, yet horticulture, strawberries, and other high-value garden crops have increased, particularly in peri-urban zones, while permanent crops and wooded areas have declined. Farm sizes have grown as the number of holdings decreased, reflecting broader trends of consolidation. This evolution demonstrates a dual agricultural model in which intensive, small-scale horticulture and livestock coexist alongside extensive arable and industrial production, shaping the physical landscape, local economies, and the Ruhr's capacity to adapt to market and ecological changes, offering both challenges and opportunities for integrated metropolitan food system planning.

Interacting Scales of Production in Milan and the Surrounding Lombardy Region

The agricultural landscape surrounding Milan presents a complex interaction between highly specialized peri urban production and broader regional agricultural patterns. While the focus here is on Milan and its immediate peri-urban areas, analyzing Lombardy as a whole provides essential context for understanding how local practices coexist within a larger, multifunctional food system. The data analyzed in this case primarily derives from the administrative limits of the Lombardy Region, allowing us to relate Milan as its functional core to the surrounding agricultural activities and explore how these practices interact within the region's complex urban and rural system. This approach does not imply that the regional government defines a functional food system; rather, it provides relevant information for planning and governance and situates Milan within a wider Functional Urban Area (FUA) (Adobati et al., 2015). Milan's peri-urban countryside is characterized by paddy fields and fodder crops for dairy, whereas surrounding provinces diversify into cereals, horticulture, viticulture, and permanent grasslands for livestock. These variations reflect not only differences in topography, soil, and climate, but also historical trajectories, governance structures, and market dynamics. Examining Milan within this multi-scalar framework highlights how local food systems are shaped by the coexistence of intensive, specialized production, emerging organic and sustainable initiatives, and long-standing livestock traditions—factors that any transition toward more diversified and environmentally sustainable forms of production cannot overlook.

According to 2023 ISTAT demographic data, the Municipality of Milan has over 1.3 million inhabitants. The former Province of Milan, now the Metropolitan City of Milan (Città Metropolitana di Milano), includes 133 municipalities to the south, east, and west of the city, with over 3 million residents. As Morandi (2005) observes, the metropolitan de facto urban area extends well beyond these boundaries, forming a continuous urbanized mesh that reaches the northern lake area toward Varese, Como, and Lecco, and the northeast toward Bergamo. This area is home to roughly 5.7 million people, and including Varese and Bergamo, the population reaches approximately 7 million. The entire Lombardy region, according to ISTAT 2023 data, has a population of 10 million.

Agriculture in Milan's peri-urban countryside mainly consists of paddy fields and fodder crops for dairy cattle. Lombardy, second only to Piemonte in rice production, concentrates rice cultivation in the southwestern

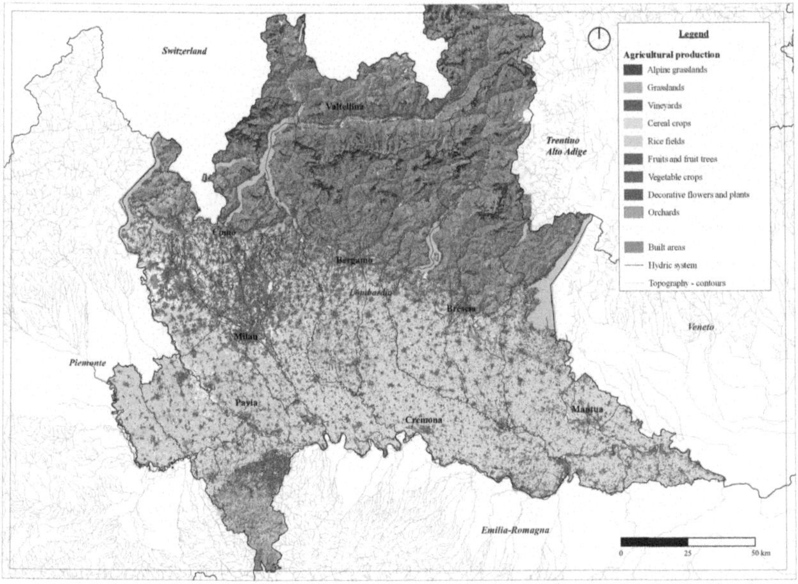

Fig. 4.3 Agricultural production in Lombardy. (Source: Elaborated by the author with data from Regione Lombardia)

territories, forming a continuous production area from Turin's outskirts to Milan and Pavia. When considering a broader Lombardy context, agricultural production diversifies. Cereals dominate near the lakes and in cities such as Bergamo and Brescia, while horticulture thrives around Mantua and the Veneto border. In pre-Alpine and Alpine areas, grazing fields and fodder crops support cattle farming. Viticulture, with IGP designation, is prominent in areas such as Oltrepò Pavese (south of the city of Pavia) and pre-Apennine hills, with additional wine production in Valtellina and the hilly zones north of Bergamo and Brescia (see Fig. 4.3).

According to the 2020 Agricultural Census, Lombardy's Total Farm Area—TA (Superficie Agricola Totale—SAT) is 1,134,384 ha, of which 83.5% is Utilized Agricultural Area—UAA (Superficie Agricola Utilizzata—SAU), and 9.2% is wooded area (WA) (see Table 4.6). The total UAA in 2020 was 946,801 ha. Although there are discrepancies[6] between ISTAT

[6]There is a discrepancy with ISTAT data. With the 2010 and 2020 numbers provided by ISTAT the net change is −2.49% of UAA. Although the number of farms is also different, the percentage is nearly similar, that is −30.10%.

Table 4.6 Type of agricultural production in Lombardy (Unit of measure: hectares and percentages)

			UAA (ha)	*% of UAA*	*% of ARA*
UAA	**Arable land (ARA)**	Cereals and grains	396,036	41.8%	54.4%
		Forage and fodder crops	249,730	26.4%	34.3%
		Industrial crops	37,165	3.9%	5.1%
		Vegetables	14,100	1.5%	1.9%
		Fallow land	12,453	1.3%	1.7%
		Legumes and potein crops	7990	0.8%	1.1%
		Other arable land crops	4464	0.5%	0.6%
		Crops under glass or high accessible cover	2490	0.3%	0.3%
		Root crops	2426	0.3%	0.3%
		Flowers and ornamental plants	1112	0.1%	0.2%
		Seeds and seedlings	700	0.1%	0.1%
		Total	**728,667**	**77.0%**	**100%**
	Permanent crops	Grapes	22,452	2.4%	64.3%
		Fresh fruit	4033	0.4%	11.5%
		Nurseries	3069	0.3%	8.8%
		Olives	2602	0.3%	7.4%
		Nuts	1113	0.1%	3.2%
		Other permanent crops	1111	0.1%	3.2%
		Berries	486	0.1%	1.4%
		Citrus fruits	63	0.0%	0.2%
		Permanent crops under glass or high accessible cover	12	0.0%	0.0%
		Total	**34,940**	**3.7%**	**100%**
	Kitchen gardens	**Total**	**300**	**0.0%**	
	Permanent grasslands	**Total**	**182,894**	**19.3%**	
	Total UAA		**946,801**	**100%**	

Source: Elaborated by the author with ISTAT data

Source: ISTAT - Dashboard Agricoltura 2024

and Eurostat regarding total TA and UAA figures, Eurostat data indicate that between 2010 and 2020, total UAA in Lombardy decreased by 1.4%, while the number of farm holdings declined by 13.9%. Data also reports that in Italy, overall UAA decreased by 6.3% (EU average −2.44%) and farm holdings dropped by 30.3% (EU average −24.79%). These trends indicate that the average farm size in Italy and Lombardy has increased, consistent with a 26.4% increase in farm size nationally since the 1980s.

Lombardy's Total Farm Area represents 47.5% of the region's total land area, and its UAA accounts for 7.6% of Italy's total UAA. Of Lombardy's UAA, 77.0% is arable land, 19.3% permanent grassland, and 3.7% permanent crops (Table 4.6). Within arable land, 54.4% is devoted to cereals and grains (41.8% of total UAA), while forage and fodder crops account for 34.3% of arable land (26.4% of total UAA). Industrial crops represent 5.1% of arable land (3.9% of total UAA), and vegetable production 1.9% of arable land (1.5% of total UAA). Figure 4.3 shows that vegetable production is concentrated mainly in Mantua, representing 4.1% of the total UAA. In highly specialized provinces like Pavia, where rice predominates, vegetables account for only 0.6% of UAA, while cereals cover 64.5% of UAA. In Mantua, 92% of UAA (153,202 ha) is arable land, and in Pavia, 91.3% (161,384 ha).

In northern Alpine areas such as Sondrio Province, permanent grasslands constitute 89% of total UAA (61,313 ha), with only 8.1% arable land, primarily for forage and fodder crops. These are well-known dairy and meat production areas. Within permanent crops in Lombardy, vineyards account for 64.3%, followed by fresh fruits (11.5%), ornamental plants (8.8%), and olive plantations (7.4%). Lombardy has 15 wine production areas organized into consortia, including Alpine vineyards in Sondrio and other valleys in Bergamo and Brescia. The region produces 5 DOCG, 21 DOC, and 15 IGT wines, which are central to the area's terroir and identity.

Some production areas, especially Pavia and Brescia, have shifted toward organic practices. In Franciacorta, renowned for DOCG sparkling wines, organic production reaches 57% (2255 ha) of vineyard area, while in Pavia, 35% (1431 ha) of vineyard area is organic (Regione Lombardia, 2019). These highly specialized areas have developed strong communication and marketing strategies, intersecting with planning and governance efforts. Such initiatives include the formation of agricultural districts (*distretti agricoli*) and biodistricts, such as the San Colombano al Lambro wine area in Pavia and the Valtellina biodistrict, all approved by Regional Decree n.436 (2017).

Lombardy, particularly around Milan, also specializes in animal husbandry, mainly dairy cattle and pig farming. Notable products include local Alpine and pre-Alpine cheeses, as well as internationally recognized cheeses such as Grana Padano, with historical significance for farming techniques and tenancy systems. As mentioned, 26.4% of total UAA is devoted to animal feed production (Table 4.6). Lombardy raises 52.1% of Italy's pigs and 27.4% of its cattle, with densities of 65.3 cattle per km^2 and

190.5 pigs per km^2, well above national averages of 18.8 cattle/km^2 and 28.9 pigs/km^2. Most livestock are concentrated in Brescia, Mantua, and Cremona.

The agricultural system of Milan and its surrounding Lombardy region demonstrates both spatial and functional differentiation, shaped by historical specialization and evolving market demands. Peri-urban areas near Milan are dominated by highly specialized rice and dairy production, while more distant provinces show a mix of cereals, horticulture, and viticulture, including high-value organic vineyards. Livestock production remains significant, supporting both local and international supply chains. Over time, farm consolidation, the expansion of organic and biodistrict initiatives, and targeted planning regulations have transformed production patterns, balancing the pressures of urban proximity with the demands of regional and global markets. This duality—between specialized peri-urban production and diversified regional agriculture—frames both the landscape and local food systems, offering insights for metropolitan planning, ecological transition, and sustainable agricultural governance.

References

Adobati, F., Ferri, V., & Pavesi, F. C. (2015). Dentro e fuori le Città metropolitane: Un'analisi della domanda di governo urbano. *Radici, Condizioni, Prospettive*, 33–46.

Diputació de Barcelona. (2015). *Informe sobre la planificación territorial y la dinámica metropolitana de Barcelona*. Diputació de Barcelona.

Handelsblatt Research Institute. (2021). *People. Nature. Space. Green infrastructure in the Ruhr metropolis.* Regionalverband Ruhr.

Instituto Nacional de Estadística (INE). (2020). *Censos y estadísticas demográficas.* INE.

Landesbetrieb Information und Technik Nordrhein-Westfalen (IT.NRW). (2020). *Bevölkerung und Fläche der kreisfreien Städte im Ruhrgebiet.* IT.NRW.

Landwirtschaftskammer Nordrhein-Westfalen. (2012). *Landwirtschaftlicher Fachbeitrag zum Regionalplan „Metropole Ruhr"*. Landwirtschaftskammer NRW.

Landwirtschaftskammer Nordrhein-Westfalen. (2019). *Zahlen und Daten zu Landwirtschaft und Gartenbau in der Metropole Ruhr.* Landwirtschaftskammer Nordrhein-Westfalen.

Morandi, C. (2005). *Milano. La grande trasformazione urbana.* Marsilio Editori.

Ostrom, E. (1990). *Governing the Commons: The Evolution of Institutions for Collective Action.* Cambridge University Press.

Regionalverband Ruhr. (2023). *Landwirtschaftliche Leitlinien*. Regionalverband Ruhr.

Regione Lombardia. (2019). *La Lombardia e i suoi vini*. Regione Lombardia.

Rocamonde, J., & Sabaté Bel, J. (2019). Paisajes Cooperativos. Una comparativa de dos paisajes industriales vinculados al río Llobregat. *Ciudades*, (22), Article 22. https://doi.org/10.24197/ciudades.22.2019.151-172

CHAPTER 5

Resource Management for Food Production

Introduction

Responsible management of natural resources for agriculture is among the pillars of the socio-ecological transition of territories that, in recent years, social and environmental movements supporting *agroecology* have pushed to advance the understanding of the role of soil, water, energy, and natural biodiversity (Gentili & Zampetti, 2021). Acknowledging the management dimension of these resources as a key aspect entails understanding that they are "collective assets" (Magnaghi, 2000; Ostrom, 1990) governed by complex policy processes that require negotiation and mediation (Janssen et al., 2006). In recent years, there has been an increase in "food-energy-water nexus policies" led by local governments in the European and North American contexts. However, in many contexts, their integration with urban agriculture has not been fully explored (Fox-Kämper et al., 2023).

This dimension introduces a territorial-based approach to provide a broader understanding of the natural elements that enable agricultural production, considering the management dimension of these finite and precious resources. It attempts to contextualize food production and farming activities in relation to other territories and systems, providing a more complex and richer framework for defining local food systems. Through specific case studies within the three analyzed contexts, this chapter focuses on mechanisms and practices for managing natural

© The Author(s), under exclusive license to Springer Nature Switzerland AG 2026

A. J. Salvador, *Metropolitan Food Systems*, Sustainable Urban Futures, https://doi.org/10.1007/978-3-032-26016-1_5

resources—such as water, soil, and seeds—as a key leverage point for achieving food security. It also takes into account historical and contested efforts to preserve these resources, drawing on vignettes and stories assembled from interviews, conversations, and secondary sources.

Preserving the Multilayered Hydric System of the Po River Valley

The Milanese case is particularly compelling due to a combination of natural, geological, and anthropogenic factors that have historically enhanced agricultural productivity. Its proximity to abundant freshwater descending from the Alps through rivers such as the Ticino, Lambro, and Adda—tributaries of the Po River—has been fundamental. This is reinforced by the presence of *risorgive* (resurgence springs), a geological condition in which pre-Alpine waters recharge aquifers in the lower plains. Human innovation has further intensified these advantages. Groundwater extraction systems such as *fontanili* (spring-fed irrigation channels) enabled advanced irrigation techniques, including *marcite* (irrigated winter meadows), allowing year-round fodder production (see Fig. 5.1). At a regional scale, from the fifteenth century onward, the construction of an extensive canal network—shaped by figures such as Leonardo da Vinci—transformed the territory's food and resource systems. Canals like the Naviglio Grande, Naviglio della Martesana, and the Villoresi Channel continue to support urban water supply and agricultural production. This multilayered hydric system has acquired multiple functions beyond productivity, including ecological, landscape, recreational, protective, and historical roles. Magnaghi (1995) defined this palimpsest as a *civilizzazione idraulica* (hydraulic civilization), a notion later central to the Lambro–Seveso–Olona strategic plan,[1] which placed the hydrological basin at the core of territorial planning.

According to the 2020 Agricultural Census, Lombardy hosts the largest irrigated area in Italy, 570,835 hectares out of 2.5 million nationwide. Of its total utilized agricultural area (1,006,985 hectares), 57% is irrigated and 68% irrigable. This abundance of water has historically underpinned agricultural prosperity, yet climate change has increasingly challenged this

[1] It refers to the strategic plan led by the Lombardy Region in 1995, titled *Bonifica, riconversione e valorizzazione ambientale del bacino dei fiumi Lambro, Seveso, Olona*, published in *Urbanistica Quaderni*, no. 2 (1995), IReR Lombardia–INU.

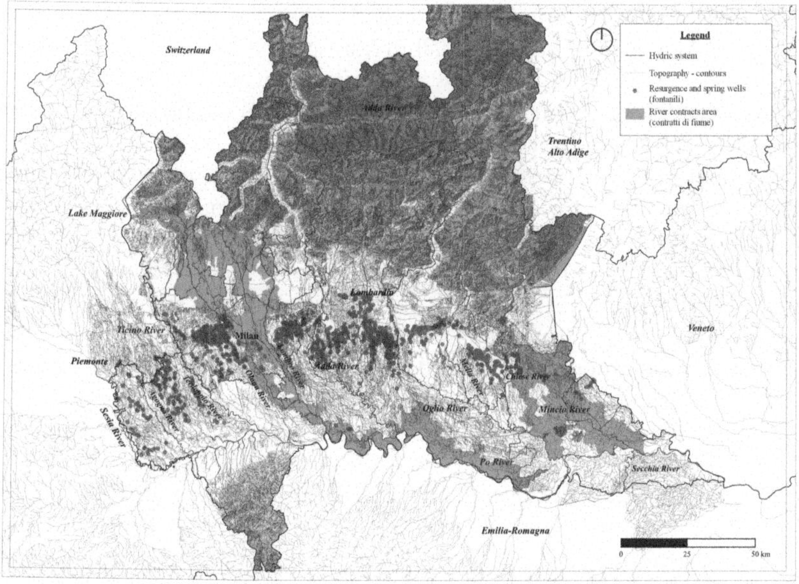

Fig. 5.1 Hydric system in Lombardy. (Source: Elaborated by the author with data from Lombardy Region)

condition. In response, farmers have adapted their practices, while administrations have introduced new governance mechanisms. During the 2022 water crisis, for instance, the Municipality of Milan convened a water policy roundtable (*tavolo dell'acqua*), deciding to lower the water levels of the Navigli canals and the Darsena basin to irrigate farmland during a critical period of the cropping cycle. This emergency response highlighted the need for stronger regional coordination on mitigation and adaptation strategies.

Hydrological data from 1951 to 2022 show a decline in total and snow precipitation, aquifer levels, and soil moisture, alongside rising average temperatures and the year 2022 marked an extreme deviation from historical averages.[2] Although 2024 recorded higher precipitation—particularly in late winter and spring—it also experienced above-average

[2] As reported by the *Sistema Nazionale per la Protezione dell'Ambiente* (SNPA), the average temperature in Lombardy throughout 2024 was +1.5 °C above the 1991–2020 average, and total precipitation was 47 % higher than the 1991–2020 average.

temperatures and extreme weather events, including floods in Milan and Lodi (15 May 2024), heavy rainfall in Varese (6–8 July, 2024), and river overflows in September. Interviews with rice farmers revealed that intense spring rains forced a shift from dry seeding to water seeding, recurring to techniques used by previous generations.

Within this broader context, the Vettabbia Valley case exemplifies growing awareness around water protection and resource management and has triggered multiple policymaking, planning and design initiatives. Located south of the Corvetto neighborhood, the valley—also known as the *Monks' Valley*—extends along a dense irrigation system and includes the medieval abbeys of Chiaravalle, Mirasole, Viboldone, historic farmhouses (*cascine*), agricultural service roads, hamlets, and remnants of natural landscapes such as woodlands, hedges, and tree rows (*siepi e filari*).

Named after the historic Vettabbia canal linking Milan to Melegnano, the area has been the focus of key interventions aimed at preserving landscape quality and sustaining agriculture. Notably, these include the Nosedo wastewater treatment plant (2003), which supplies treated water to the historic agricultural system, and the adjacent recreational park connecting Corvetto with Chiaravalle Abbey. These projects have transformed residents' perceptions of water and redefined its role in a territory long shaped by hydric infrastructures but overlooked by the city. Historically, the Vettabbia canal functioned as a sewage collector, and before reaching a state of severe contamination, its nutrient-rich waters fertilized fodder fields, establishing a metabolic relationship between the city and its surrounding countryside. (Prusicki, 2006).

The Nosedo water treatment intervention is a pilot project resulting from the cohesive strategies introduced by the Lambro–Seveso–Olona Strategic Plan—mentioned previously—which aimed to restore an integrated system of water management and territorial stewardship in this area. This strategic project represented a groundbreaking effort in Italian planning by introducing a holistic approach to environmental reclamation based on multi-level governance, active civil society engagement, and a territorial management mechanism. Within this framework, concrete actions were envisioned to regenerate the relationship between rivers and their territories, including the development of innovative governance tools centered on ecological elements, such as the *contratto di fiume* (river contract). As described by Prusicki (2006), the Nosedo project and the environmental regeneration of the *Lower Milanese countryside* enabled a rethinking of the relationship between urban and rural tissues and a re-signification of Milan's agricultural system, which had been severely threatened by aggressive speculative development during the 1970s and 1980s (Balducci & Piazza, 1981). The original proposal envisioned a comprehensive landscape and environmental intervention across the entire Vettabbia Valley, far larger than the area ultimately implemented between the landmarks of Cascina Nosedo and Chiaravalle Abbey. A further component of this vision was commissioned in 2009 by the Municipality of San Giuliano Milanese for an 818-hectare urban agricultural park.[3] Within this scenario, agriculture's economic and cultural role becomes central in countering urban expansion and other environmentally detrimental activities, such as isolated industrial zones and gravel pits embedded within the boundaries of Parco Agricolo Sud.

The Vettabbia area is particularly relevant for understanding how the agrarian landscape and economy were reduced to a generic *agricultural green space label (verde agricolo)* from political, planning, and economic perspectives. This reduction coincided with the adoption of monocultural and industrial production models (Agostini et al., 2006), through which the Milanese countryside became highly specialized and lost much of its historical complexity. Prusicki's (2006) proposal for the Vettabbia Valley emerges from this context. On the one hand, it draws from the sensitivity toward territorial assets and resources introduced by the Lambro–Seveso–Olona strategic plan; on the other, it aligns with an ambitious reinvention of Milan as a post-industrial city, supported by political vision and planning instruments. By placing farmland and agricultural production at the core of its strategies, the proposal deconstructs a segment of Parco Agricolo Sud, distinguishing it through its historical evolution, productive practices, and recent episodes in which significant portions of farmland were *sacrificed* to urban-centric development models in conflict with the agrarian landscape. The importance of this agricultural axis and its landscape qualities was also acknowledged in Andreas Kipar's 2005 Green Corridors proposal (*Raggi Verdi*) for Greater Milan, particularly RV4, which schematized a *raggio della campagna (the countryside corridor)* linked to the Pompeo Leoni urban renewal project and an open stretch of the Vettabbia irrigation ditch (*roggia*).

[3] It refers to the 2009 proposal *Parco Agricolo Urbano della Vettabbia. Progetto per la riqualificazione paesistico-ambientale della Valle della Vettabbia*, designed by Marco Prusicki and Francesca Simonetta.

The strategic proposal for the Vettabbia Valley Park is particularly relevant for discussing a scale of intervention that acknowledges the complexity of the agrarian landscape, the dynamics of farming practices, and the specific threats affecting this territory. Its agro-forestry approach is grounded in a historical analysis of agrarian development, the moments of disruption and degradation of this economy, the ecosystemic role of these areas, and the emerging challenges resulting from planning decisions and other pressures. Water is identified as the key structuring resource: The Vettabbia canal becomes the *fil rouge* around which a cohesive, multiscalar proposal is articulated. Embedded within sectorial agricultural, environmental, and water management policies, the project positions the water system as both a functional and structural element of landscape restoration, enabling agriculture to reassume a central role and re-establish its relationship with environmental, social, and economic contexts.

Water Management and the Transformation of the Llobregat River Territorial System

One of the most dramatic scenarios in Europe in early 2024 occurred in Catalunya, where devastating storms abruptly ended 2 years of drought and water scarcity. Later in the year, further intense storms affected other areas along the Spanish coast, particularly Valencia and, to a lesser extent, the Baix Llobregat area. This situation contrasted sharply with the decisions adopted in 2023 by Barcelona's metropolitan authorities to restrict water use, especially for farming activities. Barcelona's metropolitan area is home to approximately 3.3 million inhabitants and constitutes one of Europe's major tourist destinations and industrial hubs, generating intense pressure on water and other natural resources. Water-use restrictions directly affected farmers, who were among the first groups to lose access. Given the *huerta* agricultural system practiced in the area, which relies on year-round irrigation, these measures had a significant negative impact on food production. Farmers and their association, *Unió de Pagesos*, reported this decision in the media and formal press releases as one of the reasons for joining the farmer-led protests that spread across Europe in early 2024.

While the perception of this territory has shifted considerably over time, beyond the political visions that sustained these changes, it is essential to understand the area as part of a larger ecological system. In this regard, the Llobregat River plays a crucial role in supplying irrigation water, while remaining highly vulnerable to competing demands, flood risks, and the cumulative impacts of human activities on soil, ecosystems, and underground water resources. Understanding the Baix Llobregat Agricultural Park as a peri-urban agricultural space requires situating it within the broader territorial system of the Llobregat River, considering both its current functioning and its historical infrastructural transformations.

Originating in the Cadí mountain range, the river flows for 170 km before reaching the Mediterranean Sea, forming a delta of high ecological value. For centuries, agricultural practices adapted to the area's susceptibility to flooding and rising sea levels through techniques such as *estanyants (natural wetlands)*. However, the construction of two irrigation canals in the nineteenth century—the Canal de la Infanta and the Canal de la Dreta—marked a decisive economic transformation by enabling irrigated agriculture. During this period, the Delta became a major food supplier for Barcelona and a recognized exporter across Europe, earning the name *huerta de Europa*—a designation that, in retrospect, underestimated the industrialization processes[4] that would later profoundly alter the relationship between the river and its territory (Rocamonde & Sabaté Bel, 2019).

Throughout the twentieth century, the lower Llobregat plains and the Delta assumed multiple functions for Barcelona, often to the detriment of agriculture, natural ecosystems, and freshwater aquifers. A critical turning point occurred in the early and mid-twentieth century, when planning and governance frameworks framed the area as a "vacant space" (Rocamonde & Sabaté Bel, 2019), facilitating its transformation into a strategic node for urban expansion and economic development. Under this economically driven vision, the Llobregat River was heavily exploited and diverted, particularly to support the expansion of the seaport. By the 1980s, it had become one of the most polluted rivers in Europe, prompting an ambitious water reclamation program in the early 2000s aimed at improving ecological conditions through wastewater treatment plants, controlling seawater intrusion, recharging wetlands, and restoring riverine and deltaic environments (Mujeriego et al., 2008). The Land Use Plan of the Baix Llobregat Agricultural Park explicitly acknowledged the need to restore

[4]The Llobregat river structured an important system of textile production colonies that will completely alter the river's social and cultural perception but especially transform's the area's economy.

wetlands and coastal areas and to implement water treatment strategies, including the use of regenerated water for irrigation.

To fully grasp the role of resource management in this territory, it is necessary to revisit the historical processes that produced an uneven irrigation system across the Baix Llobregat plains. The current system originated in the nineteenth century as a visionary and transformative effort led by farmers, enabled by fertile delta soils and access to freshwater from the pre-Pyrenees. Today, the Llobregat irrigates approximately 78,000 hectares along its course. In the Delta, however, water allocation reflects competing priorities: 75% is destined for urban uses, 21% for industry, and only 4% for agriculture (Martí Elias, 2024).

Although not all agricultural land in the Delta is directly irrigated by the Llobregat, the system relies primarily on two canals: the Infanta Canal (1819) and the Dreta Canal (1855). The Infanta Canal represents a remarkable example of social cohesion, initiated by farmers and landowners in the absence of royal funding, and led to the creation of a still-active management entity, the *Comunidad de Regantes* (Rocamonde & Sabaté Bel, 2019). While this initiative inspired the Dreta Canal, large portions of the Delta continue to depend on groundwater wells, lacking the historical social cohesion associated with canal-based irrigation. This partly explains the prevalence of smaller plots oriented toward self-consumption, which mapping exercises show to be more vulnerable to fragmentation, land-use change, and speculation. Today, the Infanta Canal primarily serves urban water systems, with many sections either buried or disconnected from agricultural use. Current debates increasingly point toward a collective transition to drip irrigation systems (see Fig. 5.2).

Despite being a collective social, cultural, and technical achievement that transformed the Delta's economy, irrigation infrastructure has remained marginal compared to transportation and industrial infrastructures strategically located on fertile delta lands since the 1920s.[5] These include the airport, seaport, logistics hubs, highways, and railways, continuously expanded over the past century. Among the most invasive interventions was the re-routing of the river's final segment to facilitate port expansion. Field visits conducted in 2023 and 2024 reveal how this infrastructural proliferation has sealed agricultural land, fragmented the landscape, and degraded farming conditions. In many of these interstitial spaces, only informal subsistence allotment gardens remain productive.

[5] As noted by reviewing historic maps of the Institut Cartogràfic i Geològic de Catalunya.

Fig. 5.2 Analysis of the irrigation system of the Baix Llobregat agricultural park and its impact on the structure of the agricultural plots. (Source: Elaborated by the author with data and information from various sources)

Finally, understanding the lower Llobregat plains and Delta within a broader network of natural areas is essential, both from a strategic planning perspective and from a landscape ecology and biodiversity standpoint. In the early twenty-first century, proposals emerged to create a green belt (*anella verda*) adapted to Barcelona's topography and valley system (Batlle i Durany, 2000), encompassing 11 protected areas across more than 100 municipalities and totaling approximately 65,000 hectares. Within this framework, the Baix Llobregat Agricultural Park played a key role in providing spatial continuity, challenging its interpretation as a mere production platform and acknowledging its ecological and potential recreational functions.

Today, the Baix Llobregat Agricultural Park operates simultaneously as a space for food production and as part of a broader green system, governed by overlapping but fragmented institutional frameworks. As noted by staff from the Diputació de Barcelona during interviews conducted in June 2024, provincial departments manage green infrastructure and agricultural spaces separately, while water governance falls under regional authorities. This sectorial fragmentation has produced contrasting visions and remains a key challenge to be addressed through a more integrated territorial strategy.

Green Regional Planning for the Ecological Restoration of the Ruhr Area

Understanding the role of the Regional Association Ruhr (RVR) as a historical process—its evolution and current institutional form—seems essential for describing the Ruhr region, particularly in relation to planning and landscape management. Founded in 1920 as a "settlement association," RVR was initially created to guide the rapid infrastructural expansion driven by the influx of workers needed for the flourishing mining industry (Bottmeyer, 2011). As mining activities declined, RVR's mandate evolved substantially, shifting toward mitigating the economic, social, and environmental impacts of this transition. Today, the association leads regional planning and development processes and coordinates strategies among the region's cities and districts. However, its role has also been shaped by political dynamics at the state level. For instance, between 1998 and 2005, the government of North Rhine-Westphalia reduced some of RVR's planning competences (Gruehn, 2017). Despite these tensions, RVR is widely regarded as a pioneer in integrating land management with regional planning. A landmark example of its approach is the International Building Exhibition (IBA) Emscher Park (1989–1999), discussed in Chap. 4.

As part of this evolving role, RVR has also become a major landowner. Since the 1960s, the association has pursued land acquisition as a strategic tool to implement ecological restoration projects. By 2011, RVR owned approximately 17,000 hectares (Bottmeyer, 2011). Initially, acquisitions focused on large natural areas—such as woodlands and pastures near cities—but later expanded to include former industrial sites, wastelands, and areas adjacent to water bodies. This strategy enabled the creation of multifunctional public open spaces and a continuous network of green areas and cycling infrastructure, closely linked to the regeneration of a heavily polluted hydric system. Throughout these interventions, the overarching objective has been to improve residents' living conditions while enhancing the ecological quality of degraded natural resources.

Following the implementation of extensive landscape management strategies and hundreds of site-specific interventions across the Ruhr Valley, new challenges have reshaped RVR's priorities. The growing interest in renewable energy, for instance, has led to wind and solar projects, while more recent efforts have increasingly addressed agriculture and food production. As highlighted in interviews with RVR representatives in May 2023, the association has recently acknowledged its role as a farmland owner and responded through new strategies, including improved leasing conditions and longer-term contracts. This shift also reconnects with the region's pre-industrial history, in which prior to the nineteenth century, agriculture was the dominant land use.

A distinctive feature of the RVR's planning practice is its reliance on *informal* planning instruments—such as the Master Plans for the Emscher Landscape Park and the Future Emscher Valley—which are closely aligned with formal tools, including regional, comprehensive, and landscape plans (Gruehn, 2017). The Ruhr *green belt system* is one of the regional strategies established as a mechanism for the governance, planning, and management of green and open spaces through locally implemented instruments, initiatives, and projects. It functions as a continuous green infrastructure adapted to the region's polycentric morphology. Since its emergence in late nineteenth-century Great Britain, the green belt has been conceived as a regional strategy locally implemented to contain urban sprawl, protect natural land, create agricultural reserves, provide amenities for urban dwellers, and accommodate strategic infrastructure. In the Ruhr,

Zepp (2018) traces early ideas of green belts back to 1912, although their formal integration into German spatial planning occurred only in the 1960s.

Already at the end of the nineteenth century, cities such as Essen employed forms of *green planning*, notably under mayor Erich Zweigert and planner Robert Schmidt, whose contributions included the creation of the Essen city forest (Grütter, 2017). These ideas were later transferred to the regional scale when Schmidt became director of the Ruhr Coal District Settlement Association (SVR)[6] in 1920. As documented in *Grün in der Stadt Essen. Mehr als Parks und Gärten*, green areas have been embedded in local and regional strategies since the late nineteenth century and in formal planning since the early twentieth century. By 1923, SVR had already documented existing green spaces, and by 1961 their surface had increased substantially. Planning documents such as the 1927 *Generalsiedlungsplan Essen* and the 1929 *Stadtgestaltung Essen, Wirtschaftsplan* illustrate how green areas were integral to economic development strategies, including improving access to natural areas, expanding green spaces in working-class districts, and developing emblematic projects such as Grugapark and the Baldeneysee reservoir (Grütter, 2017).

The concept of *green belts* was formally introduced with the 1966 Regional Development Plan. This plan marked a turning point, responding to the decline of coal and steel industries and the availability of approximately 10,000 hectares of industrial brownfields (Dosch & Porsche, 2008). It defined seven regional green belts, later codified by letters in the 1985 Regional Open Space System and still used today (Zepp, 2018). While the designated intervention area expanded from 281 km^2 in 1966 to over 696 km^2 by 2004, significant land loss within the belts occurred due to infrastructure expansion and new settlements. Nevertheless, the regional green belt system developed into a complex spatial structure delivering a wide range of ecosystem services and becoming deeply embedded in the region's planning culture.

[6] This was the founding name of the association now known as the *Regionalverband Ruhr* (RVR). The new name was adopted in 2004, when the association regained its regional planning responsibilities. Between 1979 and 2004, the association operated under the name *Kommunalverband Ruhrgebiet* (KVR, Communal Association of the Ruhr), a period marked by a diminished political role and reduced planning responsibilities.

Importantly, the 1966 plan was the first to explicitly include agriculture as a distinct land use alongside forestry and recreation. It differentiated between fragmented agricultural areas within the core urban region—integrated into the green belt system—and larger, less fragmented agricultural areas outside it. Agriculture was particularly prevalent in the E, F, and G belts, located in municipalities such as Recklinghausen, Bochum, Witten, and Dortmund. These open spaces were conceived as continuous systems structuring dense residential, commercial, and industrial areas in relation to mobility and hydric infrastructures. The 1989–1999 IBA Emscher Park significantly reinforced the green belt system, particularly within the Ruhr's core area defined by the A2 and A40 motorways. This period also introduced the east–west green-blue corridor along the Emscher River, later consolidated in the 2005 Emscher Landscape Park Master Plan. During the 1990s, more than 23 km^2 of former industrial and residential land were reclassified as natural areas and integrated into the green belt system (Regionalverband Ruhr, 2015).

The 2010 Emscher Landscape Park Master Plan further strengthened the relationship between the green belt system, the Emscher corridor, the broader hydric network, and the region's polycentric structure. While earlier proposals were closely linked to car-oriented infrastructure, the master plan prioritized integration with cycling and pedestrian networks, reinforcing accessibility, leisure, and inter-municipal connectivity. The park now includes over 230 km of cycle paths. Of its 45,754 hectares, agriculture accounts for 17,128 hectares (37%), making it the dominant land use. The plan expanded several green belts and extended the east–west corridor along the Lippe River, combining large-scale spatial strategies with site-specific urban and landscape renewal projects. RVR's land acquisition policy was instrumental in enabling these interventions (Bottmeyer, 2011).

Agriculture gains a central role in the 2010 plan through explicit recognition, mapping, and the framing of *urban agriculture* as the predominant practice within the Emscher Landscape Park. This approach emphasizes proximity to urban areas and encourages horticultural production, direct sales, organic farming, and multifunctional farm models integrating recreational, educational, and tourism activities. The plan was developed in collaboration with the Chamber of Agriculture of North Rhine-Westphalia and other agricultural stakeholders, recognizing farms as individual economic units embedded within a larger territorial system and differentiating production morphologies across the park.

Subsequent planning instruments reinforced this open space and green infrastructure vision at the regional scale. The 2015 Ruhr Regional Plan emphasized physical continuity and introduced "municipal green corridors" (*Kommunale Grünverbindungen*), coordinated by the newly established Open Space Protection Department. In 2022, the *Green Infrastructure Charter: Ruhr Metropolis* was adopted as a policy framework and political declaration of intent (Regionalverband Ruhr, 2022), promoting connectivity, multifunctionality, and the integration of green and grey infrastructures. The charter places climate change, biodiversity, and circularity at the core of regional planning and fully acknowledges agriculture and food production as essential components of the open space system, reinforcing their role in local economies, resource management, and food systems governance.

Cascina Battivacco: Pursuing Food Sovereignty Through Seed Diversification

Cascina Battivacco is a farm located at the edge of Milan's Parco Agricolo Sud, adjacent to the Barona neighborhood. Its rice fields, colliding visually with the long gray facades of the 1960s public housing buildings of Quartiere Domus Teramo, offer one of the most striking images of the abrupt social, economic, and spatial relationships between the city and the agricultural park. This area, known as *Parco delle Risaie* (Paddy Fields Park), is one of Milan's historic rice-producing areas (see Fig. 5.3). The notion of *park* in this context emerged only a few decades ago as an attempt to better define peri-urban agricultural areas as environmental and social resources, increasingly equipped for leisure activities for nearby residents. Cascina Battivacco is a 100-hectare farm owned by Fondazione Patrimonio Ca' Granda, a non-profit foundation managing a historic public agricultural estate, and has been leased to the Fedeli family since the 1960s. The farm dates back to the twelfth century and was owned by various religious orders until 1879, when it was donated to the hospital Ospedale Maggiore. The Fedeli family has a long-standing presence in the Milanese countryside, with different branches independently managing several farms, including the well-known Cascina Resentera. This branch, represented by Angelo and Lucia, has leased Battivacco since 1965. Today, the farm is primarily managed by Cesare Fedeli, the second generation to

Fig. 5.3 *Parco delle Risaie* (Paddy Fields Park) in the Milanese countryside. (Source: Author)

lease this farmstead and the fourth generation of the family to farm in the area, having originally arrived from nearby Lodi province.

In recent years, the farm has expanded its operations and now actively cultivates around 250 hectares, including land in the Lomellina area in the province of Pavia, south of Milan. Battivacco's production focuses on rice and meat; although rice remains the primary specialization, the farm has diversified into cattle breeding to support direct sales, currently maintaining around 40 Brune cattle. Battivacco was among the first farms in the area to open a direct sales shop, offering its own products alongside selected local goods. It is also widely recognized as a pioneer of the *agriturismo lombardo* model in the 1980s, adapting agri-tourism practices from central Italy to the Lombardy and Milanese context. This diversification strategy was primarily led by Lucia Fedeli through the refurbishment of the former workers' quarters (*casa dei salariati*). Over time, the farm's activities expanded to include educational programs with schools, summer camps, and weekend initiatives, currently coordinated by Martina, Cesare's wife. Battivacco is actively involved in the initiatives promoted by Fondazione Patrimonio Ca' Granda and has become one of its flagship tenants, as well as a key stop in the itineraries curated for their Oasi *Ca' Granda* initiative.

Beyond its multifunctional model, Battivacco is particularly relevant for its role within the rice supply chain, as it has evolved into a seed producer. In 2020, Cesare Fedeli co-founded *Milano Sementi* together with geneticist Eugenio Gentinetta, a company that has rapidly gained support among farmers in Lombardy and Piemonte and has contributed to reclaiming market shares—up to 70–90% for varieties such as Carnaroli or Baldo—that were previously dominated by multinational corporations. Milano Sementi reintroduces traditional rice varieties that had disappeared due to market pressures and the standardization imposed by global seed companies. Within the framework of food sovereignty and agroecology, access to seeds represents a crucial pillar of the socio-ecological transition (Gentili & Zampetti, 2021). This activity explains the expansion of the Fedeli farm, as around 70% of its rice production is now committed to seed production. The remaining shares are sold directly at the farm shop, delivered to local rice companies, and supplied to Milan's public-school canteens as part of the municipality's local products purchasing strategy under its Food Policy. Battivacco is also a partner of Fondazione Patrimonio Ca' Granda for the foundation's rice brand, which will be discussed in Chap. 7.

Since Cesare took over the farm's management, Battivacco has progressively adopted environmentally oriented practices. Nevertheless, the farm continues to struggle with regulatory demands imposed by European and national authorities and the agri-food sector. Fedeli highlights how regulations designed for large corporations are often transferred to farmers, who must absorb the costs and responsibilities. Several sustainability-oriented investments—such as energy self-sufficiency or changes in rice drying systems—were undertaken independently and at the farm's own risk to comply with requirements imposed by large buyers and their *green* strategies.

Although the farm has significantly evolved over the past decades, Cesare Fedeli and his father still carry out farming activities that require specialized skills. He emphasizes that one of the main challenges in scaling up production is the limited access to skilled and specialized labor. For example, the annual manual weeding (*mondare*) of paddy fields requires the involvement of a specialized team brought from China, echoing practices historically performed by women across the Po Valley. Fedeli is also actively engaged in social and political organizations and expresses concern about the management of *Parco Agricolo Sud*, which he perceives as having lost its capacity to support agriculture and having become largely a political arena. As noted by Marsden (1999), the rise of multifunctional family farms has transformed farmers into "small rural businessmen," and Battivacco not only fits this description but also extends into social and political spheres, with farms becoming central actors in debates on food systems and socio-ecological transitions.

While this case highlights the agency of farms within local food systems, it also shows that in a highly politicized context farms often face challenges individually and with limited institutional support—an issue that *Fondazione Patrimonio Ca' Granda* has begun to address, and which will be further discussed in Chap. 7.

This chapter is the first of four that focus on different thematic dimensions, which the author considers essential for understanding food systems within the complexity of metropolitan agglomerations. It introduces a framework of relationships shaped by enabling resources, particularly water, and promotes a broader vision of agriculture as a form of land use that should be considered within larger natural systems—not only in terms of planning and governance, but also in terms of biodiversity and ecological interactions. A central argument of this chapter is the need to strengthen connections with other planning and governance instances, placing agriculture in dialogue with issues often treated as sectoral domains. Finally, the chapter presents the farm as a powerful lens for understanding the socio-ecological transition of food systems and territories, highlighting everyday challenges and tangible local actions that can have relevance and resonance across larger territories and scales.

References

Agostini, S., Pizzingrilli, P., & Rausa, P. (2006). *Il patrimonio rurale vernacole ai margini della metropoli. Analisi di quattro insediamenti nell'area metropolitana milanese*. Libreria CLUP.

Balducci, A., & Piazza, M. (1981). *Dal Parco Sud al cemento armato. Politica urbanistica e strategie immobiliari nell'area milanese.*

Batlle i Durany, E. (2000). L'anella verda de Barcelona: La matriu ecològica metropolitana. *Àrea: Revista de Debats Territorials, 8*. https://futur.upc.edu/1640065

Bottmeyer, M. (2011). Land Management of Former Industrial Landscapes in The Economic Metropolis Ruhr. *Bridging the Gap Between Cultures*. FIG Working Week 2011.

Dosch, F., & Porsche, L. (2008). Grüne Potenziale unter blauem Himmel. Neue Zugänge zur Flächenrevitalisierung und Freiraumentwicklung im Ruhrgebiet. *Informationen Zur Raumentwicklung, 9*, 609–625.

Fox-Kämper, R., Kirby, C. K., Specht, K., Cohen, N., Ilieva, R., Caputo, S., Schoen, V., Hawes, J. K., Ponizy, L., & Béchet, B. (2023). The role of urban agriculture in food-energy-water nexus policies: Insights from Europe and the U.S. *Landscape and Urban Planning, 239*, 104848. https://doi.org/10.1016/j.landurbplan.2023.104848

Gentili, A., & Zampetti, G. (Eds.). (2021). *Agroecologia Circolare. Dal campo alla tavola. Coltivare biodiversità e innovazione.*

Gruehn, D. (2017). Regional Planning and Projects in the Ruhr Region (Germany). In M. Yokohari, A. Murakami, Y. Hara, & K. Tsuchiya (Eds.), *Sustainable Landscape Planning in Selected Urban Regions* (pp. 215–225). Springer Japan. https://doi.org/10.1007/978-4-431-56445-4_18

Grütter, H. T. (Ed.). (2017). *Grün in der Stadt Essen. Mehr als parks und gärten.* Klartext.

Janssen, M. A., Goosen, H., & Omtzigt, N. (2006). A simple mediation and negotiation support tool for water management in the Netherlands. *Landscape and Urban Planning, 78*(1–2), 71–84. https://doi.org/10.1016/j.landurbplan.2005.05.005

Magnaghi, A. (Ed.). (1995). *Bonifica, riconversione e valorizzazione ambientale del bacino dei fiumi Lambro, Seveso e Olona. Linee orientative per un progetto integrato.* Istituto Nazionale di Urbanistica.

Magnaghi, A. (2000). *Il progetto locale.* Bollati Boringhieri.

Marsden, T. (1999). Rural Futures: The Consumption Countryside and its Regulation. *Sociologia Ruralis, 39*(4), 501–526. https://doi.org/10.1111/1467-9523.00121

Martí Elias, J. (2024). Subsol del delta: L'aigua oculta i els equilibris ecosistèmics al tram final del Llobregat. *Seminario Internacional de Investigación en Urbanismo*, (16), Article 16. https://doi.org/10.5821/siiu.12931

Mujeriego, R., Compte, J., Cazurra, T., & Gullón, M. (2008). The water reclamation and reuse project of El Prat de Llobregat, Barcelona, Spain. *Water Science and Technology: A Journal of the International Association on Water Pollution Research, 57*, 567–574. https://doi.org/10.2166/wst.2008.177

Ostrom, E. (1990). *Governing the Commons: The Evolution of Institutions for Collective Action.* Cambridge University Press.

Prusicki, M. (2006). Area Sud Milano. Unos scenario strategico di riqualificazione paesisistica del Basso Milanese. In *Landscape Opportunities. La gestione paesistica delle trasformazioni territoriali. Complessità territoriale e valorizzazione del paesaggio* (p. 131).

Regionalverband Ruhr. (2015). *Fachliche Grundlage „Regionale Grünzüge" zum Regionalplan Ruhr.* Regionalverband Ruhr.

Regionalverband Ruhr. (2022). *Green Infrastructure Charter. Ruhr Metropolis*. Regionalverband Ruhr.

Rocamonde, J., & Sabaté Bel, J. (2019). Paisajes Cooperativos. Una comparativa de dos paisajes industriales vinculados al río Llobregat. *Ciudades*, (22), Article 22. https://doi.org/10.24197/ciudades.22.2019.151-172

Zepp, H. (2018). Regional green belts in the Ruhr region. A planning concept revisited in view of ecosystem services. *Erdkunde*, *72*. https://doi.org/10.3112/erdkunde.2018.01.01

CHAPTER 6

Emerging Farming Styles for Sustainable Practices

Introduction

A farm business model, also defined as *farming style*, can be understood as a set of strategic notions imbued with territorial identity that guides farmers' activities, services, and decision-making processes (Van der Ploeg et al., 2009). These styles manifest in the way agriculture is practiced, including the farm's productive orientation, intensity, and diversification strategies. Since the 1990s, the re-localization of food markets has seen the role of the farmer bifurcate between being a "small rural businessman" and a "local environmental manager" Marsden (1999). As extensively described in Chap. 2, multifunctionality is a term widely employed across disciplines, especially in agronomic studies, to describe an economic model that emerged at the end of the twentieth century, aimed at diversifying farm economies through multiple activities (Marsden, 1999; van der Ploeg et al., 2009). In peri-urban areas, farms have increasingly adopted multifunctional business models, combining on-site services for leisure and recreation with environmental contributions, such as landscape management for public landowners, and social benefits, such as allotment gardens or community-supported agriculture (CSA) schemes (Lohrberg et al., 2015). All these supplementary activities represent additional sources of income, providing financial stability through revenue diversification.

Farming styles operate at both the individual farm scale and the territorial level. While a farm's model shapes its immediate production, services,

© The Author(s), under exclusive license to Springer Nature Switzerland AG 2026

A. J. Salvador, *Metropolitan Food Systems*, Sustainable Urban Futures, https://doi.org/10.1007/978-3-032-26016-1_6

and activities (Swift & Anderson, 1994), collectively, farms influence the cultural identity of a region and produce recognizable landscapes and agricultural systems (Smeding & de Snoo, 2003). This occurs through coordinated practices, shared techniques, and responses to urban demands, shaping the types of crops grown, retail strategies, and production methods implemented.

Family farming is the dominant agricultural model in Europe. According to the Food and Agriculture Organization (FAO)[1], 97% of farms (around 12 million holdings) are considered to be family run. EU[2] data considers 93% of farms (9.1 million holdings) to be categorized as family farms in 2020, with an average size of 11.3 hectares, compared to 102.2 hectares for non-family farms. Although non-family farms represent only 7% of holdings, they control 39% of agricultural land and generate disproportionately higher economic output. Family farming, as defined by FAO, is primarily characterized by its reliance on family labor and is commonly associated with smaller-scale, diversified business models, short supply chains, strong ties to local communities, and greater sensitivity to landscape and biodiversity. At the same time, it faces shared challenges, including complex administrative requirements, the prevalence of small and medium-sized holdings, and the need for generational renewal, with farmers under 40 often prioritized in support schemes.

This approach places the farm holding, and within it the family farm, as the fundamental unit of analysis in agrarian studies, recognizing it as both an economic and socio-ecological entity. Acknowledging the relevance of farming styles leads to an analysis of the conditions necessary for farmers to sustain their practices, innovate, and participate in socio-ecological transition. This chapter explores these conditions in detail, focusing on the essential resources for production, the diversity of business models, and illustrative examples of farms that embody the evolving strategies of peri-urban agriculture. Field visits, interviews, and engagement with local markets and direct-sales points form the empirical foundation for this analysis, highlighting how individual farms both shape and respond to their territorial and cultural contexts.

[1] Information gathered from: https://www.fao.org/family-farming/countries/eur/en/#:~:text=Family%20farming%20is%20the%20most,including%20farms%20of%20all%20sizes. Accessed on 13/01/2025.

[2] Information gathered from: https://ec.europa.eu/eurostat/statistics-explained/index.php?title=Agriculture_statistics_-_family_farming_in_the_EU. Data accessed on 13/01/2025.

Preserving Farming Models in the Baix Llobregat Agricultural Park

The Baix Llobregat Agricultural Park is a territory that lends itself to studying how, collectively, individual agricultural practices are capable of forming a system and producing a social fabric that can self-organize to become a relevant territorial actor. This case is examined through the process of the park's formation, the role that farmers played in its creation and consolidation, and the way in which agricultural practice becomes the common denominator. At the same time, it explores the role of the park's managing body, which, by recognizing specific types of production and agriculture, seeks not only to preserve them but also to stimulate and enhance them. In this way, agricultural practice becomes a shared objective and a value to be safeguarded. This case also shows how other types of production may be ignored and marginalized within these processes. However, they play an important role in keeping productive areas that are increasingly of less interest to the very model that is initially defended.

In 1977, the *Unió de Pagesos* (Peasants' Union of Catalonia) published *Salvem el Pla!*, a manifesto calling for the protection of the agricultural lands of the lower Llobregat River plain within Barcelona's metropolitan area. These lands played a crucial role in supplying food to Barcelona and surrounding municipalities but were increasingly threatened by urban expansion, industrial development, and strategic infrastructures such as the airport and the port. Following the post-war period, rapid urban growth and planning visions that overlooked the territory's landscape, social fabric, and productive value placed this historic agrarian area at risk (Montasell i Dorda, 2010). Since then, the Unió de Pagesos has been a central actor in shaping the Baix Llobregat Agricultural Park, providing political advocacy, social mobilization, and agronomic expertise. Founded clandestinely in 1974 in opposition to Franco's regime, the union emerged as a successor to the *Unió de Rabassaires*, a syndicate of sharecroppers established in 1922 and dissolved in 1939, which had fostered strong cooperative traditions (Planas, 2022). Today, the Unió de Pagesos is Catalonia's most influential farmers' union, with over 6000 members, and the figure of the *pagès* (peasant and farmer) remains deeply embedded in Catalan cultural and territorial identity.

The Parc Agrari del Baix Llobregat was formally established in 1998 through a local development and management plan and currently protects

3489 hectares[3] of agricultural land. Governed by a consortium including 14 municipalities, regional and provincial authorities, and the Unió de Pagesos, the park operates within a landscape heavily fragmented by infrastructure (see Fig. 6.1), most notably Barcelona's international airport. Agricultural production is dominated by fruit and vegetable farming on family-run smallholdings (*minifundios*) and medium-sized plots. Products are distributed through farmers' markets, on-farm shops (*agrobotigues*), and a dedicated pavilion at Mercabarna for local producers and cooperatives. Emblematic products include artichokes, asparagus, *calçots* (a type of Catalan green onion), cherries, and the *Pota Blava* chicken, several of which hold IGP certification.

Fig. 6.1 Rail infrastructure and agricultural plots in the Baix Llobregat. (Source: Author)

[3] This figure refers to the area officially designated as the agricultural park. According to the 2014 revision of the park's Special Plan (*Revisió del Pla Especial de Protecció i Millora del Parc Agrari del Baix Llobregat*), the park's functional area covers 2911 hectares. The revision expanded the park's total surface from 3323 to 3348 hectares. More recently, a 2023 presentation by the park's manager, Helena Perxacs, reported 1924 hectares of actively farmed land, while the 2014 plan classifies 2426 hectares as protected agricultural land within the park (*Clau 24b – Rústic protegit de valor agrícola del Parc Agrari*).

The park is managed through the planning and management plans: *Plan de Gestión y Desarrollo* and a *Plan Especial de Ordenación*, coordinated by the consortium. Its initiatives address key challenges of peri-urban agriculture, including land abandonment, access to land and water, pest management, urban pressure, collective marketing, and farmers' capacity building. Despite these efforts, structural difficulties persist, particularly regarding the transition to environmentally oriented production systems, with only a limited number of certified organic producers currently operating in the park, which will be further discussed in Chap. 8 together with the efforts led by the park's management entity.

A Culturally Dominant Farming Style

Interviews with farmers consistently point to a relatively stable business model that has prevailed for decades. This model typically consists of small farms of 4–6 hectares dedicated to horticultural and fruit production, usually managed by a couple and close family members. Fieldwork occupies most of the week, while 1 or 2 days are reserved for harvesting, preparation, and distribution through cooperatives, markets, shops, or, more recently, ecological purchasing groups (*grups de consum ecològics*). External labor is used only to address seasonal needs.

In recent years, many farmers have diversified their distribution channels by selling directly through on-farm shops (*agrobotigues*), weekly farmers' markets (*mercats de pagès*), permanent stalls in municipal markets, or direct sales to consumers. This shift reflects both initiatives promoted by the agricultural park's management entity and growing consumer interest in local food. The expansion of Mercabarna—Barcelona's wholesale market—and the creation of a section dedicated to local producers through cooperatives has offered an additional outlet (see Fig. 6.2). However, despite its proximity, some farmers perceive Mercabarna as commercially distant, requiring crossing Barcelona, and instead prioritize markets in nearby regions such as Girona.

Farmers have also increasingly engaged in food processing, either independently or through external partners, often as a way to valorize products rejected for aesthetic reasons. Although the park's Land Use Plan originally envisioned shared facilities (*agropols*) for storage, processing, and packaging, these spaces were never implemented despite clear demand. The Unió de Pagesos is currently advocating for their reintroduction. Similar collective infrastructures have been supported elsewhere by the

Fig. 6.2 Mercabarna's section allocated exclusively for local cooperatives. (Source: Author)

Diputació de Barcelona, such as communal slaughterhouses, while in the Baix Llobregat these activities are largely self-organized or mediated through the cooperatives in the towns of Sant Boi de Llobregat or El Prat de Llobregat. The need for shared facilities is exacerbated by limited on-farm space and restrictions on new construction within the park.

Field visits and interviews revealed how individual and collective initiatives enable these distribution mechanisms, for instance the case of *pagès* Albert Bou Samsó illustrates this dynamic. Operating a 5-hectare horticultural farm with over 250 years of history—reduced in size due to a highway expansion—he played a key role in establishing a weekly farmers' market in Colònia Güell through collaboration with the local municipality. His farming calendar is organized around market attendance and a stable customer base, and he coordinates crop planning with other farmers to diversify supply and stabilize output. While he previously engaged in small-scale processing, he now collaborates with the Fundación Espigoladors, which transforms surplus or aesthetically imperfect produce with social and environmental aims. Mercabarna remains his last-resort outlet through the cooperative wing dedicated to local producers.

These practices highlight that sustaining agriculture in this peri-urban context requires farmers' engagement beyond cultivation alone. It depends on collective organization, coordination of production calendars, participation in associations and cooperatives, and continuous interaction with local administrations and the park's management entity. Distribution and cooperation thus emerge as both economic strategies and political practices are essential to maintaining agricultural activity in the Baix Llobregat.

Abandonment of Agricultural Land: Subsistence Agriculture and Conflicting Farming Scales

The average age of farmers in Baix Llobregat is 60, while in El Maresme it is slightly lower, at 57; both figures are close to the national and Catalan averages of 61. The share of farmers under 45 years of age reaches 20% in El Maresme and 16% in Baix Llobregat, exceeding the national average of 13%. Although El Maresme displays a relatively higher presence of younger farmers, the proportion of female-led farms remains below the national level, with only 14% managed by women. In Baix Llobregat, 15% of farms are female led, compared to 24% in Catalonia and 29% nationally.

Among the challenges connected with the predominantly family-based nature of these farms—where most operators are landowners and the average age exceeds 60—generational renewal emerges as a critical issue for ensuring the continuity of agricultural activities. This challenge affects European agriculture more broadly and is no exception in this territory. From the perspective of the Baix Llobregat Agricultural Park's management entity, generational renewal is a major concern, both in relation to the practices adopted by aging landowners and the heightened risk of land abandonment. While the narrative of the park as a project built on cooperation and cultural identity remains relevant, it is increasingly under strain as farming struggles to attract younger generations and as established farmers face difficulties in engaging with evolving practices and broader socio-ecological expectations. Generational renewal, increasingly approached through a gender perspective, also emerged as a recurring concern in interviews with farmers and other local actors in Barcelona. Across Europe, it is identified as one of the main challenges facing agriculture, shaped by economic constraints, administrative burdens, the demanding nature of farm work, and younger generations' heightened concern for climate change (Borda et al., 2023).

In the Baix Llobregat Agricultural Park, the impacts of this transition become particularly visible through the coexistence—and tension—between different scales, trajectories, and orientations of agricultural production. Fieldwork and interviews reveal five distinct farming scales, differentiated by production goals, plot size, tenure regimes, and social profiles. These scales increasingly generate friction over access to land and resources, agricultural techniques, and their role in shaping the park's identity. The first scale corresponds to professional family farmers described before (*pagesos*), cultivating smallholdings of approximately 5–7 hectares and producing fruit and vegetables for nearby markets, direct sales, and the wholesale market Mercabarna. The second scale originates in mid-twentieth-century subsistence practices in Gavà and Castelldefels, characterized by highly fragmented plots (around 1500 m^2) used for self-consumption and small-scale commercial production. Shaped by limited access to irrigation infrastructure, these areas have undergone continuous transformation as some plots have been abandoned or converted into leisure gardens, while others have been consolidated by larger producers. Their ambiguous status, neither fully commercial nor formally regulated, places them in competition with professional farming.

The third scale involves the subdivision of farmland into informal allotment gardens rented to private individuals, often as an income strategy for aging farmers. Although prohibited by park regulations, this practice persists due to its significantly higher economic returns compared to renting land to other farmers. While the Land Use Plan foresaw municipally managed allotment gardens in transitional areas, formal examples remain rare. The fourth scale refers to subsistence family orchards occupying neglected or infrastructurally fragmented land, particularly near highways and along the Llobregat River. Often stigmatized and lacking formal recognition, these practices represent some of the few remaining forms of cultivation in degraded areas. As Pirro and Anguelovski (2017) argue, despite their growing presence, subsistence gardens are largely excluded from planning frameworks.

The fifth scale reflects the gradual entry of corporate farming through land acquisition and consolidation strategies[4]. Although not new, this process is more pronounced in Baix Llobregat than in other Catalan counties,

[4]To access the newspaper article: https://www.elllobregat.com/noticia/26954/lluis-m-estruch/a-contra-corriente:-el-grupo-ametller-se-planta-en-el-parc-agrari-en-tiempos-de-sequia.html

driven by strategic proximity to the wholesale market, the port, airport, and other logistics infrastructures. The increasing share of farms operated by legal entities raises concerns about shifting social and economic dynamics and the erosion of traditional farming structures.

This multi-scalar overview highlights agriculture as both economic and cultural practice shaped by uneven transitions. While initiatives such as land banks and testing grounds for new farmers represent important institutional responses, they struggle to address the structural aging of the farming population and the resulting land abandonment. Recognizing non-commercial and subsistence practices as part of the productive landscape is essential, as their exclusion not only fragments agricultural land but also limits access to resources, knowledge, and environmental support mechanisms.

Towards Multifunctional Agriculture in the Ruhr Metropolis

Farming in the Ruhr Metropolis mainly follows three dominant business models: low-cost specialization, differentiation, and diversification (Pölling et al., 2016). Low-cost specialization takes advantage of the region's proximity to urban markets, allowing farms to intensify production while minimizing transport costs. Differentiation emphasizes high-quality, niche, or value-added products, often supported by innovative production techniques, marketing strategies, or participatory schemes such as community supported agriculture (CSA) and rent-a-field initiatives. Diversification extends beyond traditional farming, combining multiple revenue streams that integrate agricultural and non-agricultural activities, from tourism and social services to landscape management. Together, these models illustrate how Ruhr farms negotiate the pressures and opportunities of a densely urbanized, polycentric region, balancing economic viability with social engagement and struggling commitment to ecological stewardship.

As revealed by NRW Chamber of Agriculture data, among the three governmental regions (*Regierungsbezirke*) within the Ruhr Metropolis, Münster has both the largest agricultural area (389,810 hectares) and the highest number of farm holdings (10,350), despite not having the largest total territorial extent. Between 2005 and 2020, however, Münster experienced the most pronounced decline in utilized agricultural area (UAA), with a reduction of 4.70%, exceeding that of the other two regions. Over the same period, Arnsberg recorded the steepest decline in the number of farm holdings (–34.70%), a trend broadly consistent with the national average (–32.66%).

Farm size dynamics further illustrate structural change. Across the three regions, average farm size increased from 28.8 hectares in 2005 to 40.6 hectares in 2020 (+40.82%). At the national level, average farm size reached 63.1 hectares in 2020, representing a comparable increase of 44.51% since 2005. According to the NRW Chamber of Agriculture 2019 report (Landwirtschaftskammer Nordrhein-Westfalen, 2019), the average farm size in the wider Ruhr Metropolis was approximately 40 hectares in 2019, with notable intra-regional variation. A comparison of Eurostat data in the state of North Rhine-Westphalia also indicates a significant increase between 2010 and 2020 in the number of farms engaging in additional agricultural and non-agricultural activities. At the national level, this share rose from 30.80% to 40.98%, a trend also reflected in the regions of Düsseldorf, Münster, and Arnsberg. The increase is most pronounced in Münster (+37.83%), followed by Düsseldorf (+26.83%) and Arnsberg (+17.07%), pointing to differentiated dynamics between the more urbanized core and the more rural hinterland.

Across all regions, renewable energy production represents the most widespread additional activity, particularly in Münster, where it involves 48.37% of farms with gainful activities, and aligns with both national trends and proximity to infrastructure and markets. In contrast, Arnsberg—characterized by more extensive and peripheral agricultural conditions—shows a stronger orientation toward forestry. Agricultural contract work also plays a significant role across regions, reflecting economies of scale and inter-farm cooperation.

Overall, these patterns suggest that diversification has become a key strategy across the Ruhr Metropolis, while the specific forms it takes vary spatially: Energy production and contract work are more prominent in the core and its immediate surroundings, whereas forestry-related activities remain more relevant in the hinterland. Renewable energy, in particular, often functions less as an independent income source than as a means of reducing operational costs within increasingly constrained farming systems.

Livestock farming also emerges as a key component of multifunctional systems in North Rhine–Westphalia. Beyond its productive function, the presence of animals enables direct marketing of products such as milk, eggs, and meat, and underpins a range of complementary activities involving both farming and non-farming animals. Livestock also shapes land-use patterns and farming calendars, as significant areas are allocated to grazing and fodder production, reinforcing integrated crop–livestock systems. The presence of horses and other equines plays a more prominent social and recreational role, especially in Düsseldorf and Arnsberg. Overall, livestock farming contributes not only to economic diversification but also to landscape management and social functions, reinforcing its role within adaptive and multifunctional agricultural systems. According to the 2012 North Rhine–Westphalia Chamber of Agriculture, 71% of the farms' agricultural turnover derives from animal husbandry, while 29% comes from the cultivation of market crops.

This analysis characterizes agriculture in the Ruhr area not only as multifunctional—reflected in the diversification of farm incomes—but also as a provider of a wide range of resources and services. Beyond food production, agriculture contributes social and environmental benefits, including the cultivation of renewable raw materials. Moreover, farming in the region is described as highly adaptive, having continuously responded to pressures such as urban expansion, insecure land tenure, spatial constraints, and re-naturalization policies. Finally, the report underscores that the quality and diversity of open spaces cannot be preserved, planned, or managed without a functioning agricultural system, emphasizing agriculture's central role in food and raw material production, its requirements as an economic activity, and its often-overlooked capacity to deliver unpaid social and environmental services (Landwirtschaftskammer Nordrhein-Westfalen, 2012).

Business Models and Farming Transition

As reported in the 2020 Agricultural Census, the transition toward organic farming remains a pending challenge, both in terms of agricultural surface and the number of farm holdings. In a comparison of the number of farms adopting organic practices in 2010 and 2020, an increase can be observed in all three regions as well as at the national level. These figures remain low when compared to other European countries and fall well short of the EU target of reaching 30% organic agriculture by 2030. The data also reveal significant territorial differences in agricultural production within the Ruhr Metropolis, particularly in the distribution of ecological farming. Within the metropolitan core, the share of organic farming is generally lower than in areas characterized by more extensive agricultural systems in southern districts such as Hagen (8.8%) and Ennepe-Ruhr (7.5%). At the same time, it exceeds the values observed in several northern production areas, including the districts of Unna (2.5%) and Wesel (2.7%), as well as the municipality of Hamm (3.5%). Among core municipalities, Essen stands out with a relatively high share of organic farming (6.3%), followed by Dortmund (5.5%), both exceeding the state average of 4.2%.

According to the 2020 agricultural census, female farm managers represent around 10% of total holdings in the Ruhr Metropolis and approximately 7% of the utilized agricultural area (UAA), figures closely aligned with the national average. While the absolute number of female-managed farms has declined since 2005, their relative share has increased due to a faster decrease in male-managed farms. Farmers are predominantly older: 35–38% fall within the 55–64 age group, followed by those aged 45–54, together accounting for nearly two-thirds of all farmers. Female managers under 25 years old are virtually absent, reflecting a broader national trend of limited generational renewal. Family labor remains central to farming structures, with over 80% of farms relying primarily on family workforce. Part-time farming is also widespread, as more than half of farm holders engage in additional gainful activities alongside agriculture. Interviews confirm that full-time farming is often associated with transitions toward ecological practices and multifunctional business models. The Schulte-Tigges educational farm, discussed in this section, illustrates how a full-time commitment can enable diversified activities such as community-supported agriculture and educational services.

One of the main contributions of this research is to frame farms as bottom-up practices that interact with governance, planning, and land-management strategies across scales, revealing how agriculture shapes land use and territorial functions. The two case studies on the outskirts of Dortmund represent a new generation of farms that have diversified their business models to ensure economic viability while maintaining a strong social orientation. Analyzing these models provides insight into contemporary farming trajectories, the challenges farmers face, and the tools available to planners and policymakers to support the socio-ecological transition of agrarian territories.

Dickhöfer Organic Farm

The Dickhöfer farm is a 70-hectare family-owned organic farm located in Waltrop, in the district of Recklinghausen, north of Dortmund, of which

15 hectares are owned and 55 leased. As part of the CoProGrün project[5], the farm collaborates with the Emscher river authority (*Emschergenossenschaft*) to manage 21 hectares of restored floodplain landscape in the Erlebnispark Emscherland[6] through grazing with a rare Red Mountain cattle breed (*Rotes Höhenvieh*). This arrangement integrates agricultural production with ecological restoration and landscape management and provides the farmers with additional income.

The farm dates back to the 1600s and is currently managed by Jan and Eva Dickhöfer, supported by their two children. Jan took over the farm around 20 years ago, making farming his full-time job after previously working in carpentry. The farm has been organically certified since 2016, following a long transition marked by economic difficulties during the first years. Crop production focuses on corn, wheat, spelt, and Champagne rye, including ancient grain varieties supplied directly to local bakeries.

In recent years, the farm has diversified its activities, integrating free-range egg production, small-scale beef production, and the rental of family gardens. Free-ranging chickens and a mobile coop are a visible feature of the farm, and eggs are sold through a self-service vending machine. Beef production is limited to around 28 cattle, with 2 to 3 animals slaughtered annually, yielding approximately 1500 kilograms of meat. Slaughtering is carried out on-site using a mobile unit, and meat is sold directly through the farm and online in small quantities, with an emphasis on reducing waste by offering a wide range of cuts.

The garden rental project, led by Eva, has become one of the farm's core activities and emerged from the CoProGrün initiative. Now in its fifth year, it involves around 30 families cultivating small plots within a 2000 m^2 area. The project combines food production with training in organic practices and biodynamic principles, contributing to biodiversity and community engagement (see Fig. 6.3). Adjacent land has also been

[5] Project funded by the Federal Ministry of Education and Research (Bundesministerium für Bildung und Forschung – BMBF) as part of the **Innovative Municipalities** program (*Kommunen innovativ*). The applied research project, **CoProGrün – Co-produced Green Spaces as Sustainable Municipal Infrastructure**, focuses on the Ruhr Metropolis' Green Belt F and aims to promote a sustainable development model by directly involving farmers and other civil society groups

[6] This project was a collaborative effort involving the Emscher River authority, the cities of Castrop-Rauxel, Recklinghausen, Herne, and Herten, and the Regionalverband Ruhr – RVR. It was primarily funded through the European Regional Development Fund – ERDF.

Fig. 6.3 Family gardens at the Dickhöfer organic farm. (Source: Author)

made available to young farmers experimenting with vegetable production under tunnels.

Cooperation is central to the farm's model, ranging from equipment sharing with neighboring farms to partnerships with the Emscher river authority. Jan is actively involved in farmers' associations, organic certification bodies, and Regionalwert[7] AG, a citizen-led investment model supporting regional food sovereignty, while Eva is engaged in social and political initiatives. Although most labor is family-based, the farm employs part-time support for farm and administrative work. By opening the farm to direct interaction with customers and the wider community, Jan and Eva aim to raise awareness of the realities of food production, while acknowledging current economic pressures, climate-related challenges, and the difficulties faced by new generations of farmers, particularly in accessing land.

The Dickhöfer farm exemplifies a diversified, multifunctional farming model that combines ecological production, social engagement, and

[7] More information can be accessed at https://regionalwert-muensterland.de/. Last visited on 17/03/2025.

landscape stewardship, while facing common challenges related to climate change, land access, and economic pressures on consumers.

Lernbauernhof Schulte-Tigges

While cooperation characterizes the Dickhöfer organic farm, Lernbauernhof Schulte-Tigges is defined by a strong ethos of solidarity and education. The farm covers 26 hectares on the northern outskirts of Dortmund, of which 8 hectares are directly farmed by Elmar, his wife Miriam, 3 employees, and volunteers. The remaining land is leased to a nearby organic farmer under long-term (10-year) contracts. The farm has been in Elmar's family since 1621. After several years working abroad as geographers, Elmar and Miriam decided to take over the family farm. Initially managed on a part-time basis, the transition to full-time farming marked a deliberate shift toward a socially oriented business model, placing education, solidarity, and community engagement at its core.

This transition included a gradual move to organic farming, with Bioland certification granted in July 2024. Production focuses on vegetables and legumes for direct marketing, requiring less land than the farm owns and allowing the remaining land to be leased as an additional income source. A central component of this model is the establishment of a Community-Supported Agriculture scheme (*Solidarische Landwirtschaft—SoLa Wi*), complemented by an expanding range of educational activities. These began with farm visits, workshops, and interactions with animals and later evolved into a permanent daycare center for 20 children aged 3 to 5. Obtaining institutional recognition required navigating lengthy administrative procedures, reflecting the lack of precedent for such hybrid agricultural-educational models in the area. The farm also hosts regular school visits and seasonal activities such as summer camps, making it a highly active site involving participants of different ages and backgrounds. Intergenerational continuity remains present, with Elmar's father providing ongoing support, particularly through his farming knowledge.

The 8 hectares directly managed by the farm are spatially organized to support multifunctional use. Approximately 2 hectares are occupied by the farmstead, including residential buildings, barns, stables, storage facilities, educational spaces, and modular daycare units with direct street access (see Fig. 6.4). Four hectares of grassland are dedicated to feeding farm animals, while the remaining 2 hectares are used for SoLaWi vegetable production. The Kümper Heide SoLaWi, established in 2013, emerged in

Fig. 6.4 Location and layout of the Lernbauernhof Schulte-Tigges' main activities. (Source: Elaborated by the author)

response to demand from Dortmund residents seeking a partner farmer and has grown from an initial group of 40 participants to approximately 200 households. The scheme is organized around 110 annual shares, renewed each year through a collective assembly in which producers and consumers jointly plan the agricultural season and commit financially through a solidarity-based bidding system. Each share was valued at €103 per month in 2024, with higher contributions by some members compensating for those with fewer economic resources. Distribution is organized through clustered collection points to minimize transport time and emissions, with most members living within a 30 km radius of the farm.

Environmentally responsible practices are embedded throughout the farm's operations. Vegetable production takes place in well-maintained tunnels supported by flowering strips and hedges to enhance biodiversity. Drip irrigation systems reduce water consumption, and photovoltaic installations supply renewable energy. Knowledge-sharing and cooperation extend beyond the farm, including support for other farmers and involvement in a local Food Council. The educational dimension provides an important source of economic stability and reinforces the farm's social mission.

Lernbauernhof Schulte-Tigges exemplifies a socially oriented and multifunctional peri-urban farming model in the Ruhr Metropolis, where agricultural production, education, and community-based food systems are tightly linked. At the same time, this case reveals some of the most common challenges to peri-urban agriculture, including climate uncertainty, pressures on soil quality, access to land, and the challenges for transitioning to sustainable farming models. This case also offers a concrete lens through which to understand both the potential and the structural constraints of multifunctional agriculture in metropolitan contexts.

Framing the Agricultural System of the Po River Valley

Italian agriculture is highly specialized and acknowledged worldwide, primarily due to over 300 food protected designation of origin (PDO) and protected geographical indication (PGI) products and more than 500 wine products, which establish a direct relationship between farming, food, and territory. Italian agriculture caters both to national and global markets. In 2022, the agri-food sector reached a value of 550 billion euros, corresponding to 15% of the Italian economy, and employed 1.6 million people.

This decades-long process of agricultural industrialization and specialization has also led to a cultural process of building a strong identity within territories of production (Barbera, 2021), while simultaneously creating a fragile economy based on monocultures, with detrimental environmental impacts (Nifosi, 2023) and a production system heavily reliant on an exploited foreign workforce (Rondot, 2022). Although this specialization originally emerged from geographical and historical features linked to local peasant traditions and their capacity to shape complex agricultural landscapes, it gradually evolved into industrialized and market-oriented forms of production, carrying significant social, environmental, and economic consequences (Barbera, 2023). In the context of climate change and increasing environmental pressures observed over the past decade, these highly specialized, geographically certified production systems face substantial risks, challenging the economic structures that sustain them

(Privitera et al., 2025). Wine producers, among others, have been at the forefront of reporting drastic changes and implementing adaptive strategies (Will Media, 2025), although such challenges extend well beyond viticulture.

The agriculture of the Po River Valley forms part of this specialized system, in which the Milanese countryside is particularly acknowledged for rice production and dairy farming. Different scales and scopes of production enable this output, but as in the rest of Italy, family farming remains the most common type, representing 96.8% of all farms. According to the 2020 agricultural census, the average farm size in Lombardy is 21.8 hectares, with 41.8% of the utilized agricultural area (UAA) dedicated to cereals and grains. Within the province of Milan, 57.6% of the total utilized agricultural area (38,147 hectares) is allocated to cereals and grains, 23.3% (15,486 hectares) to forage, and 0.8% (509 hectares) to vegetable production. Inside the Municipality of Milan, 47.3% of the 5548 hectares are allocated to cereals, 18.2% to forage, and 0.6% to vegetables. Although the area devoted to vegetables is limited, 6.1% of farms in Lombardy participate in vegetable production, representing 8.6% of arable land farms.

While there is a gradual transition toward less environmentally harmful practices, as in the other cases, this remains a major challenge. Between 2010 and 2020, the number of farm holdings and UAA increased by 123.3% and 128.5% respectively in Lombardy, and Italy leads Europe with 29.6% of all organic farms, totaling 72,460 holdings (6.4% of national farms) and 1,817,630 hectares (15.1% of UAA). Organic farms are larger on average (25.1 hectares) than non-organic ones (9.4 hectares), a trend similar only in Germany among the cases studied.

Introducing New Farming Models Challenging Specialization

Within Milan and particularly in Parco Agricolo Sud, multifunctionality has been embraced by numerous farms and promoted by public institutions, including municipalities and land management entities—as a strategy to foster relationships between urban and rural areas, and between producers and consumers. For instance, dedicated farmers' markets within the city for local producers or promoting the activities offered by the farms. Here, landscapes assume a multifunctional role, not merely for agricultural production but as spaces capable of delivering social, cultural, and environmental benefits (Caravaggi et al., 2021), underpinning the "*patto città-campagna*" (city-countryside pact) concept, a principle advocated by (Magnaghi, 2000).

According to the 2020 Italian agricultural census, 14% of farms in Lombardy (6347 holdings) engage in activities economically integrated into the farm. The most common are touristic and educational initiatives (35%), encompassing both structured agritourism with lodging and meals and simpler programs like workshops and school visits. Food transformation accounts for 29%, including direct marketing on-farm and participation in farmers' markets. Contracting activities, either for other farms or municipalities—rank third, and renewable energy production, mainly for operational cost reduction, is fourth. Interviews revealed that farms operating within protected built-heritage structures often struggle to implement energy-reduction strategies, as such measures are frequently incompatible with heritage conservation regulations (e.g., tiled roofs that prevent the installation of solar panels). Data from 2013 and 2016 show trends toward multifunctionality, with a slight decline in most categories except renewable energy, though recent interviews indicate incremental diversification.

The Milanese experience is distinctive for its rich architectural heritage and large agricultural structures, which can be repurposed for multifunctional uses. *Cascine* (farmsteads) are reprogrammed for direct sales, agritourism, camping, and educational activities (see Fig. 6.5), often involving farmers, families, and staff. Maintaining heritage buildings—what Agostini et al. (2006) termed the "cascina problem"—has been a central challenge, yet many farms have refurbished spaces to accommodate multifunctional business models. Multifunctionality has been crucial in preserving this heritage, as farmers reinvest in refurbishing structures for new uses, while municipalities and institutions such as Parco Agricolo Sud have introduced strategies to promote public engagement. For example, the *Guida alle aziende agricole* lists farms offering public activities, and initiatives such as ZiGò in Zibido San Giacomo map cycling itineraries that highlight farms and facilitate visits.

The following cases were selected to exemplify the ongoing adaptation of agricultural business models toward multifunctionality, heritage preservation, and the strengthening of rural-urban linkages. Together, they illustrate how individual farm activities and business strategies can collectively shape territories. On the one hand, these cases show how farms reshape perceptions of and engagement with agrarian spaces; on the other, they reveal how farms increasingly operate as infrastructures for leisure and

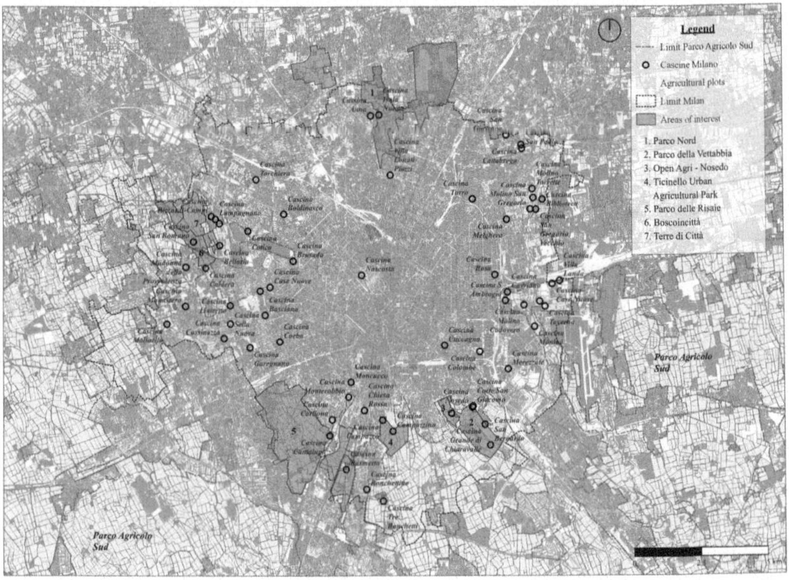

Fig. 6.5 Municipally-owned farmhouses and peri-urban agrarian spaces. (Source: Elaborated by the author)

education (e.g., Cascina Femegro), disrupt multinational-dominated production systems (Cascina Battivacco, discussed in Chap. 4), and challenge monocultural practices (Podere Ronchetto). The Oasi Ca' Granda case, discussed in Chap. 7, further demonstrates how resilient business models can be embedded within long-term management strategies that reconcile the economic needs of farms with the environmental objectives of an institutional landowner.

Cascina Femegro

Cascina Femegro is a historic, family-owned farm located within Parco Agricolo Sud, in the municipality of Zibido San Giacomo, managing approximately 65 hectares of owned land plus additional leased farmland. The farm specializes in rice and dairy production, maintaining over 350 head of cattle, including around 150 dairy cows. While most of the land is dedicated to fodder for feeding livestock, a smaller portion is used for rice cultivation sold through a direct sales point. The farm is managed by second- and third-generation family members, with operations reflecting both agricultural production and multifunctional business strategies.[8] The farm historically relied on cooperatives and large companies for product distribution, but the 2009 "milk crisis" marked a turning point, prompting the third generation to adopt a more diversified business model focused on milk transformation into cheese, yogurt, ice cream, and other dairy products. This strategy included creating permanent direct-sales spaces and introducing local artisanal products, demonstrating a shift from traditional supply chains toward more autonomous and territorialized production.

Over the years, Cascina Femegro has embraced multifunctionality as a central component of its business model. Educational and recreational activities, once occasional, now occur nearly every weekend. The farm's courtyards have become public spaces, attracting families and school groups, while additional animals were introduced to enhance visitor engagement. The family has adapted roles to accommodate these changes, managing public interaction, hosting activities, and leading communication efforts through social media. A recent initiative converted former staff housing into a bed and breakfast. Given its proximity to hospitals, it primarily targets families of long-term hospital patients from outside Milan, rather than conventional tourists.

The farm's long history, dating back to the fifteenth century, provides both a cultural and functional framework for these adaptations. Originally a moated castle and classified[9] as a "*cascina fortificata*," the property evolved into a farm by the eighteenth century and features a double-courtyard structure with additional buildings supporting agricultural production. Continuous adaptation of the architectural heritage has enabled multifunctional uses, including automated milking systems, educational programs, and direct-sales infrastructure, illustrating the close interrelation between heritage conservation and contemporary farm management.

Chiara, representing the third generation, emphasizes that multifunctionality emerged more as a necessity than as a voluntary choice, reflecting broader economic pressures on agriculture. Cascina Femegro exemplifies a *textbook case* of multifunctional farming in Lombardy, demonstrating how experimental and incremental strategies—supported by land

[8] The visit and interview were held on 25/03/2022.

[9] This refers to the document *Repertorio dei beni storico-architettonici e ambientali*, an annex to the municipality's main planning instrument.

ownership and intergenerational commitment—allow farms to diversify economic activity, engage the public, and reinforce territorial food systems. The farm actively participates in initiatives by Parco Agricolo Sud and the municipality to promote local agriculture and food, although many institutional programs, such as park product labeling, have lost momentum. Most communication, marketing, and project management are now farmer-led, illustrating both the autonomy and the responsibilities embedded in multifunctional farm management.

Cascina Femegro also provides insight into consumer interactions and territorial dynamics. Direct observation of purchasing trends, relocation patterns, and customer engagement highlights the farm's role as a critical node connecting producers, consumers, and the urban-rural interface. In this sense, the farm functions not only as a production site but also as a social and educational infrastructure, mediating between the countryside and the urban population of Milan.

Podere Ronchetto

Podere Ronchetto is a small, peri-urban farm of 7.8 hectares within Parco Agricolo Sud, leased from Fondazione Patrimonio Ca' Granda. Despite its limited size, the farm benefits from significant historical and built infrastructure, dating back to the fifteenth century and located in the *borgo agricolo* (rural hamlet) of Ronchetto delle Rane. The farm's history reflects the evolution of Milanese countryside properties, transitioning from noble estates to multifunctional agricultural spaces. Since 2016, Francesco Bossi and his family have managed the farm, focusing on bio-certified production of seasonal vegetables, rice, corn, wheat for flour, and, more recently, floriculture. The farm collaborates with nearby producers to enhance biodiversity and collectively strengthen local production networks. Products are sold directly to consumers through an on-site shop (see Fig. 6.6), community-supported agriculture (CSA) schemes, and weekly markets, establishing strong local consumer relationships. The farm's shop also serves as a reference point for visitors exploring Parco Agricolo Sud and initiatives like Oasi Ca' Granda, integrating the farm into broader territorial itineraries.

Podere Ronchetto integrates multifunctional activities beyond agricultural production. A social-oriented project engages young migrants in job training, some of whom have become permanent workers. The farm leverages its large courtyard, resulting from nineteenth-century building

Fig. 6.6 Shop and publicly accessible facilities at Podere Ronchetto. (Source: Author)

demolitions, to host a year-round campsite (*agricampeggio*) for visitors across Europe. Additionally, refurbished buildings accommodate a small bed-and-breakfast, supporting family involvement in the farm and creating new revenue streams. These initiatives highlight the farm's ability to combine production, social engagement, and public interaction.

Francesco Bossi actively participates in governance structures, serving as a representative for farmers in Parco Agricolo Sud and collaborating with family producers' associations such as the Distretto Agricolo Milanese (DAM). He emphasizes the challenges of navigating the park's governance, regulatory frameworks, and technical limitations, particularly regarding sustainable infrastructure such as solar panels. The maintenance of historic farm buildings is a recurring concern, with several structures still unused due to high costs, despite the Bossi family's investment in refurbishment and adaptive use.

Podere Ronchetto illustrates the experimental and adaptive nature of modern peri-urban farming. Its bio-certified production and innovative business model, including CSA, social engagement, and tourism-like activities, demonstrate how small farms can leverage proximity to urban centers and historical infrastructure to diversify economic activities. The farm also highlights broader systemic challenges, such as climate adaptation, technical support, and navigating multiscale governance structures, underscoring the complex role of peri-urban farms in Milan's agricultural and social landscape.

Across the three contexts, farm business models illustrate a diversity of practices and demonstrate adaptability in responding to emerging demands and pressing challenges, while also acting as active agents in the socio-ecological transition of food systems and communities. The selected cases are regarded as units of a larger system undergoing active transformation.

In Barcelona, small-scale, family-run farms integrate high-value horticultural production, proximity to markets, and cultural significance, delivering a business model that has sustained local food production for at least a century. However, this model is increasingly at risk, as emerging and contrasting farming approaches begin to replace the historical one. This case also highlights the role of local institutions, particularly the park's management entity, in attracting new generations and supporting commercial activities through various initiatives. The Ruhr Metropolis demonstrates a combination of low-cost specialization, differentiation, and diversification, with farms incorporating non-agricultural activities, renewable energy production, and livestock management as part of multifunctional strategies. The farms presented as case studies exemplify a model that strongly relies on land security and long-term commitment, often evolving into full-time operations. Issues of land ownership and stability will be further discussed in Chap. 7. In Milan, the peri-urban countryside is characterized by intensive rice and fodder production, yet farms like Cascina Femegro and Podere Ronchetto showcase integrated approaches that combine diversified production, food transformation, direct sales, and recreational and hospitality activities, while emphasizing the cultural and social potential of establishing direct relationships with consumers and institutions.

Together, these cases reveal how multifunctionality, encompassing economic diversification, ecological stewardship, social engagement, and heritage preservation, shapes contemporary farm business models, enabling agricultural systems to respond to urban pressures, environmental challenges, and generational change, while strengthening rural-urban linkages.

References

Agostini, S., Pizzingrilli, P., & Rausa, P. (2006). *Il patrimonio rurale vernacole ai margini della metropoli. Analisi di quattro insediamenti nell'area metropolitana milanese*. Libreria CLUP.

Barbera, G. (2021). *Il giardino del Mediterraneo. Storie e paesaggi da Omero all'Antropocene*. Il Saggiatore.

Barbera, G. (2023). *Agrumi. Una storia del mondo*. Il Saggiatore.

Borda, Á. J., Sárvári, B., & Balogh, J. M. (2023). Generation change in agriculture: A Systematic review of the literature. *Economies, 11*(5), 5. https://doi.org/10.3390/economies11050129

Caravaggi, L., Lanzani, A., & Longo, A. (2021). Nuovi parchi agro-sociali. Infrastrutture di cittadinanza nei territori periurbani. In *Ricomporre i divari. Politiche e progetti territoriali contro le disuguglianze e per la transizione ecologica* (pp. 271–282). Il Mulino.

Landwirtschaftskammer Nordrhein-Westfalen. (2012). *Landwirtschaftlicher Fachbeitrag zum Regionalplan „Metropole Ruhr"*. Landwirtschaftskammer NRW.

Landwirtschaftskammer Nordrhein-Westfalen. (2019). *Zahlen und Daten zu Landwirtschaft und Gartenbau in der Metropole Ruhr*. Landwirtschaftskammer Nordrhein-Westfalen.

Lohrberg, F., Licka, L., Scazzosi, L., & Timpe, A. (Eds.). (2015). *Urban agriculture Europe*. Jovis.

Magnaghi, A. (2000). *Il progetto locale*. Bollati Boringhieri.

Marsden, T. (1999). Rural futures: The consumption countryside and its regulation. *Sociologia Ruralis, 39*(4), 501–526. https://doi.org/10.1111/1467-9523.00121

Montasell i Dorda, J. (2010). El parque agrario del Baix Llobregat: Un paisaje cultural. *Identidades: territorio, cultura, patrimonio, 2*, 30–39. https://doi.org/10.5821/identidades.8794

Nifosi, C. (2023). Regolare il rurale. In *Ecologie Rurali. Pratiche e forme della coesistenza* (pp. 51–68). Lettera Ventidue.

Pirro, C., & Anguelovski, I. (2017). Farming the urban fringes of Barcelona: Competing visions of nature and the contestation of a partial sustainability fix. *Geoforum, 82*, 53–65. https://doi.org/10.1016/j.geoforum.2017.03.023

Planas, J. (2022). La Unión de Rabassaires y el cooperativismo (1922–1936). *Ayer. Revista de Historia Contemporánea, 125*(1), 1. https://doi.org/10.55509/ayer/125-2022-09

Pölling, B., Mergenthaler, M., & Lorleberg, W. (2016). Professional urban agriculture and its characteristic business models in Metropolis Ruhr, Germany. *Land Use Policy, 58*, 366–379. https://doi.org/10.1016/j.landusepol.2016.05.036

Privitera, D., Graziano, T., & Polizzi di Sorrentino, E. (2025). New regions in a glass. Embedding landscape in the production of wine narratives. *Landscape Research, 50*(1), 74–88. https://doi.org/10.1080/01426397.2024.2367471

Rondot, C. (2022). *Abitare l'opacita. Gli spazi rurali di Borgo Mezzanone.* Lettera Ventidue.

Smeding, F. W., & de Snoo, G. R. (2003). A concept of food-web structure in organic arable farming systems. *Landscape and Urban Planning, 65*(4), 219–236. https://doi.org/10.1016/S0169-2046(03)00058-6

Swift, M. J., & Anderson, J. M. (1994). Biodiversity and ecosystem function in agricultural systems. In E.-D. Schulze & H. A. Mooney (Eds.), *Biodiversity and ecosystem function* (pp. 15–41). Springer. https://doi.org/10.1007/978-3-642-58001-7_2

Van der Ploeg, J. D., Laurent, C., Blondeau, F., & Bonnafous, P. (2009). Farm diversity, classification schemes and multifunctionality. *Journal of Environmental Management, Multifunctional Agriculture - From Farm Diagnosis to Farm Design and Institutional Innovation, 90*, S124–S131. https://doi.org/10.1016/j.jenvman.2008.11.022

Will Media (Director). (2025). *Gradi. Il vino italiano ai tempi del cambiamento climatico* [Documentary]. Will Media.

CHAPTER 7

Land Ownership, Emerging Actors, and Management Strategies

Introduction

Land tenure regimes encompass the legal, social, and cultural dimensions that define how land is owned, managed, and accessed by incoming generations. As historically discussed by authors such as Sereni (1961) and more recently by Sevilla Buitrago (2022), land tenure regimes are intrinsically linked to the enclosure and management of territories, reflecting social structures, power relations, and governance models, and collectively shaping how landscapes are produced and inhabited. This set of conditions, often described as a "bundle of rights" (Baysse-Lainé & Perrin, 2018), strongly influences farming styles and business models, particularly for farmers who lease land or access farmland for the first time. Land tenure has emerged throughout this research as a key factor shaping farmers' economic viability. Several authors (e.g., Beingessner et al., 2023; Calo et al., 2021; Rotz et al., 2019) highlight that land ownership is a major barrier to agroecology, due to the concentration of land in capitalist economies and the limited access for small-scale farmers. As Otsuka et al. (1992) demonstrated, tenancy arrangements and conditions imposed by landowners can hinder transitions toward environmentally sustainable farming.

In recent years, academic and policy interest has increasingly focused on contracts as instruments to incentivize farmers and land managers to deliver environmental public goods, as exemplified by initiatives such as

© The Author(s), under exclusive license to Springer Nature Switzerland AG 2026

A. J. Salvador, *Metropolitan Food Systems*, Sustainable Urban Futures, https://doi.org/10.1007/978-3-032-26016-1_7

the Contracts 2.0 project involving institutions from 13 countries (European Commission, 2023). In parallel, local governments have begun to reassess the role of publicly owned agricultural land as a strategic and often undervalued resource for cities, both in terms of food production and territorial management.

This dimension examines specific cases within each context to understand the land management models implemented by public institutions and large institutional landowners. It addresses both virtuous examples and more contentious situations, where public or large-scale landownership lacks a coherent management framework or long-term strategy for land and associated infrastructures. The analysis draws on interviews, policy documents, and secondary sources to provide an in-depth understanding of these management models and their implications for farming practices and territorial dynamics.

Land Ownership and Tenure in Barcelona Province

According to Spain's 2020 Agricultural Census, land tenure in the Baix Llobregat county is characterized by a strong mismatch between ownership and the utilization of agricultural land. Of the 543 farm holdings, 61.88% (n = 336) are owned by the farmer. However, owned land accounts for only 25.86% of the Utilized Agricultural Area (UAA), corresponding to 1039 hectares. In contrast, 56.13% of the UAA (2255 hectares) is cultivated under leasing regimes (see Table 7.1). This reflects a structural

Table 7.1 Structure of farms by tenure (Unit of measure: % of holdings)

	Property % farm holdings	*Property % UAA*	*Lease % UAA*	*Communal % UAA*	*Other % UAA*
Baix Llobregat	61.88	25.86	56.13	14.96	3.06
Maresme	61.54	32.31	65.65	0.67	1.37
Penedès	74.61	39.3	36.14	0.27	24.29
Anoia	83.33	48.68	41.22	0.15	9.96
Barcelona	**67.69**	**31.9**	**61.34**	**0.54**	**6.22**
Catalunya	**80.72**	**38.46**	**55.19**	**1.64**	**4.71**
Spain	**86.07**	**51.75**	**37.42**	**8.07**	**2.76**

Source: Elaborated by the author with data from the 2020 Agricultural Census by the Instituto Nacional de Estadística

pattern whereby farmers tend to own smaller farms—generally below 10 hectares—while leasing larger plots. This configuration closely resembles that of El Maresme, both in the proportion of owned holdings and in the relationship between ownership and total UAA. El Maresme also shows a notable absence of commons-oriented tenure regimes, a pattern that is similarly uncommon across Barcelona province and Catalonia more broadly. By contrast, areas with different agricultural orientations, such as Penedès or Anoia, display significantly higher levels of ownership. In Anoia, ownership exceeds 80% of farm holdings and approaches 50% of the UAA, figures that align more closely with the Catalan and national averages.

At the Catalan scale, tenant farming predominates, particularly in Barcelona province and along coastal counties such as El Maresme and Baix Llobregat, while the national average of leased land stands at 37%. In this respect, Penedès and Anoia are more representative of national trends, as leased land does not surpass owned land. These differences reflect broader distinctions between peri-urban and non-peri-urban agricultural systems in terms of production types, farm size, land tenure arrangements, and management models. Although not shown here, peri-urban areas such as Baix Llobregat and El Maresme also register a higher share of farms operating as legal entities or commercial enterprises.

Urban pressure, planning policies, and marketing strategies further accentuate these differences. In peri-urban contexts, proximity to consumer markets—including wholesale facilities—favors production systems oriented toward fresh vegetables and fruit, which generally require limited transformation. In contrast, more specialized and intensive agricultural systems depend on larger land bases, more complex processing infrastructures, and distinct land management arrangements.

Approaches to Mitigating Land Abandonment

As discussed in Chap. 6, among the challenges associated with the predominantly family-based nature of these farms—where most farmers own their land and the average age exceeds 60—is the need for generational renewal to ensure the continuity of agricultural activity. This issue affects European agriculture more broadly and is particularly acute in this territory. From the perspective of the Baix Llobregat Agricultural Park's management entity, generational turnover is a critical concern, both in relation to the practices adopted by aging landowners and the growing risk of land abandonment. In response, the park's consortium, together with the *Diputació de Barcelona* (Barcelona provincial government), has begun experimenting with land bank schemes that enable young farmers to access land for training and practice. A more ambitious initiative promoted by the provincial government involves pairing farmers approaching retirement with younger farmers through an apprenticeship-like model, seeking to recreate a form of *family-style* succession in which knowledge, practices, and land are gradually transferred. This institutional effort focuses on identifying interested parties and addressing the legal conditions required to facilitate land transfer under tenancy regimes. The initiative has made significant progress in the emerging Parc Rural de Montserrat.

The agricultural park consortium also collaborates with the Institut Les Salines to train a new generation of farmers and has sought to link this educational pathway to the land bank initiative in order to reactivate abandoned land. However, in practice, some landowners opt to informally subdivide plots into small gardens, as explained in the previous chapter, a trend criticized by the Unió de Pagesos. The union supports land banks as a structured solution and has proposed additional regulatory strategies across different scales of production.

One initiative initially conceived as bold and particularly promising, explicitly aimed at generational renewal through an agroecological and social approach, is the Cooperativa Central del Parc Agrari del Baix Llobregat. Operating on 5.5 fragmented hectares of previously abandoned land, the cooperative brought together new farmers committed to principles of sustainability, justice, solidarity, and equality, with a particular focus on engaging vulnerable and fragile groups. As noted in an interview with one of the group's coordinators, establishing such an unconventional model proved professionally challenging in a context characterized by a strong and fixed cultural understanding of both the farmer's role and the farming business model. Despite several successful initial years, the cooperative ceased its activity in 2023 after 5 years of operation.

More broadly, interviews with farmers highlight the need to move beyond rigid cultural definitions of agricultural activity and to rethink family farming in more open and inclusive terms—across gender, nationality, business models, and farming practices—as well as to expand the understanding of collectiveness beyond formal farmers' associations. While this vision challenges the narrative discussed in the previous chapter, it also underscores the need for evolution.

Institutional Landowners and Land Management Strategies in the Ruhr Metropolis

Regarding land tenure, across all three regions—Düsseldorf, Münster, and Arnsberg—and at the national level, more than 80% of farm holdings are operated by landowners, with Arnsberg showing the highest share at 86.6%. However, when tenure is analyzed in relation to the total Utilized Agricultural Area (UAA), the proportion of land farmed by owners drops significantly, falling below 50% in all cases. In Düsseldorf, only 35% of the UAA is farmed by landowners, closely aligning with the national average of 38.2%, while Münster and Arnsberg show slightly higher figures at 48.8% and 44.3%, respectively (see Table 7.2).

Within the core area of the Ruhr Metropolis, between 60% and 80% of agricultural land is leased, while across the wider RVR territory approximately 55% of farmland is not owned by the farmer. In this context, land ownership emerges as a critical issue, particularly due to widespread speculative leasing practices, often based on 1-year contracts. This concern also emerged during interviews with RVR staff, and it proved central to understanding the conditions required to provide stability for farmers, especially for those seeking to transition towards more ecological farming practices.

Table 7.2 Tenure of farm holdings in 2020 (Unit of measure: holdings and hectares)

	Total holdings	Owner	Share of total holdings %	Tenant	Common land	Shared or other tenure
Düsseldorf	4820	3970	82.4%	3600	0	410
Münster	10,350	8280	80.0%	7090	0	420
Arnsberg	5890	5100	86.6%	4360	0	750
Germany	**262,560**	**233,550**	**89.0%**	**195,200**	**0**	**30,630**
	Total UAA	**Owner**	**Share of total UAA %**	**Tenant**	**Common land**	**Shared or other tenure**
Düsseldorf	219,510	76,890	35.0%	140,170	0	2450
Münster	389,810	190,350	48.8%	196,650	0	2810
Arnsberg	246,250	109,200	44.3%	130,080	0	6970
Germany	**16,578,460**	**6,336,430**	**38.2%**	**9,970,560**	**0**	**271,480**

Source: Elaborated by the author with Eurostat data

These tenure dynamics are closely linked to the region's industrial legacy. Due to historical mining and industrial activities, a significant share of agricultural land remains under the ownership of large institutional actors, including companies such as RAG Montan Industries and ThyssenKrupp Liegenschaften, as well as water management authorities (Emschergenossenschaft and Lippeverband), churches, and foundations. This ownership structure continues to shape leasing conditions and farmers' long-term capacity to invest in land stewardship and sustainable production systems.

Due to its industrial legacy and the governance structures shaped by a highly specialized territorial organization, the Ruhr Metropolis still reflects patterns of landownership rooted in its industrial past. Mining companies such as Ruhrkohle AG transitioned into sectors like real estate to manage extensive land portfolios, while other majors also include the Regionalverband Ruhr association (RVR). As described by Bottmeyer (2011), RVR has long relied on land acquisition as a strategy for landscape protection and environmental transition, particularly in relation to brownfield regeneration, park development, the conversion of former mining sites, and the consolidation of a continuous green infrastructure system across the region.

During interviews conducted in May 2023 and July 2024 with representatives of RVR, initially focusing on the implementation of the CoProGrün project, the question of land ownership emerged and so did the leasing conditions established by RVR to their farmland tenants. CoProGrün was a multidimensional applied research project carried out between 2017 and 2021, aimed at co-producing sustainable green spaces within the F Green Belt–Eastern Emscher Valley, covering municipalities such as Dortmund and Castrop-Rauxel. The project explicitly explored the role of urban agriculture and farming practices within a broader green infrastructure framework, emphasizing the co-creation of new spatial typologies through collaboration with farmers (Christenn & Timpe, 2019). Funded by the Federal Ministry of Education and Research (BMBF) under the Kommunen innovativ program and coordinated by RWTH Aachen University, the project involved RVR, the University of Applied Sciences of South Westphalia, the North Rhine-Westphalia Chamber of Agriculture, and the local planning collective Die Urbanisten.

Although CoProGrün is not analyzed in detail here, interviews revealed its importance in providing RVR with direct insight into the structural challenges faced by farmers—particularly the role of landownership in shaping economic stability, investment capacity, and long-term planning. While RVR has historically promoted regional planning as a tool to protect agricultural land and prevent speculation, the project highlighted the need to address these issues explicitly from the perspective of a landowner.

In June 2023, RVR published a set of guidelines for leasing and managing its agricultural land (*Landwirtschaftliche Leitlinien*), marking a formal commitment to improving leasing conditions and aligning land management with regional biodiversity and sustainability strategies (Regionalverband Ruhr, 2023). The document explicitly acknowledges RVR's responsibility as one of the region's largest landowners and links this role to national and regional planning frameworks. Importantly, it identifies food production and food security as central objectives of agriculture in the Ruhr Metropolis, thereby clarifying the types of practices and production systems RVR intends to support.

The guidelines highlight ongoing challenges such as continued land consumption, the predominance of tenant farming, part-time agricultural activity, and the expansion of industrial crops not oriented toward food production—issues previously discussed in Chaps. 3 and 5. These conditions contribute to uncertainty and discourage farmers from investing in their holdings or transitioning toward environmentally sustainable practices. RVR currently owns approximately 1300 hectares of farmland distributed across around 200 lease agreements, comprising 780 hectares of permanent grassland and 310 hectares of arable land. Most leases are renewed annually unless terminated with 6 months' notice, a system that has generated instability for farmers but also provides RVR with flexibility, particularly for reallocating land to environmental compensation measures. Farm plot sizes vary considerably, with 45% smaller than one hectare.

From an institutional perspective, the development of these guidelines underscored the need for stronger internal coordination among RVR's divisions, including environment and green infrastructure, regional planning, and agriculture and forestry. This coordination is essential for integrating the guidelines into binding planning instruments and operational management tools. In recent years, the RVR Ruhr Grün division has assumed responsibility for designing new lease agreements, ensuring alignment with local ordinances, landscape plans, and compensation strategies that integrate agricultural practices—an approach further illustrated in the case of the City of Bochum, to be discussed in the following section. At the European level, initiatives such as the Contracts 2.0 project (European

Commission, 2023) similarly emphasize the role of improved leasing conditions in enabling socio-ecological transitions in agriculture.

While RVR has long managed forests, protected areas, and recreational spaces, the active governance of agricultural land represents a relatively new but significant expansion of its mandate. The six guidelines articulate a vision of agriculture as a socially, culturally, and environmentally valuable activity, rooted in sustainable food production and biodiversity conservation. They promote longer-term leasing arrangements—particularly to support farmers transitioning to organic production—alongside the protection of agricultural land from infrastructure development. The guidelines also emphasize the integration of farming into environmental compensation measures, the encouragement of biodiversity-enhancing practices, support for regional food systems and producer-consumer networks, and sustained investment in environmental education and capacity building. In many respects, this approach parallels strategies adopted by entities such as the Baix Llobregat Agricultural Park, which combines land management with marketing, education, and support for ecological transition. RVR's evolving role as planner, coordinator, and landowner positions it uniquely to address agriculture from multiple perspectives and align it with broader environmental and planning agendas. However, a key challenge remains the effective integration of local municipalities—both in governance and land management—alongside other major institutional landowners operating within the Ruhr Metropolis.

Compensation Measures to Avoid Farmland Consumption

Farmland consumption due to environmental compensation measures has emerged as a recurring concern in both interviews and official reports, such as RVR's agricultural guidelines (Regionalverband Ruhr, 2023). A related issue concerns the role of local governments, which operate within regional planning frameworks yet often implement strategies independently. This dynamic is examined through interviews with representatives of the municipalities of Bottrop and Bochum, focusing on Bottrop's involvement in the KuLaRuhr project and Bochum's Production Integrated Compensation (PIK) project, which integrates compensation measures into farmland leased to farmers. Both cases illustrate how local authorities mediate between regional planning strategies, landowners, and farming practices.

In Bottrop, the City Planning Office highlighted the challenges ecological compensation poses even within urban regeneration projects. Although the municipality does not have a formal strategy for its agricultural land, it regularly acquires farmland to meet project-related needs and compensation requirements, leasing approximately 85 hectares to farmers under 1-year agreements. These short-term leases provide municipal flexibility but create uncertainty for tenants. Bottrop is committed to preserving farmland north of the city, a historically fertile area dominated by a few large, politically active commercial farmers. Urban densification strategies are promoted to reduce farmland consumption, yet environmental compensation demands for new developments require available land. Concurrently, the city is pursuing ambitious post-industrial projects, including the redevelopment of former mining sites such as Welheimer WasserMark, as well as initiatives to halve household carbon emissions and address energy poverty.

Bochum has adopted a more integrative approach. Through collaboration between the municipality, municipal farmers, and the Westphalian Cultural Landscape Foundation (*Stiftung Westfälische Kulturlandschaft*), compensation measures are directly embedded within agricultural production to enhance biodiversity, soil quality, and preserve the rural cultural landscape. The 10-year PIK project, launched in 2021, encompasses 40 hectares of municipal farmland, building on a 2019 pilot at Bockholt Hof (see Fig. 7.1). Eleven farmers now manage 20 sites, with parcels ranging from 0.25 to 5.3 hectares, primarily in the city's southern districts near the Ruhr River and Schleusen Canal. Farmers voluntarily participate and receive financial compensation alongside longer leases, extended from 1 to 5 years. Compensation is calculated based on intervention type and the number of eco-points generated.

The project integrates measures such as prohibiting fertilizers and pesticides, establishing perennial flower strips, leaving fallow land, practicing crop rotation, and creating pollinator habitats. It is actively monitored for ecological impact, including the population of wild bees. Beyond biodiversity, the interventions have aesthetic and educational value, transforming monotonous farmland into visually and socially meaningful landscapes. The project has generated 1.2 million eco-points as of July 2024,[1] demonstrating both scale and complexity growth. According to Alexandra

[1] The interview at the Municipality of Bottrop was held on July 24, 2024, with Christina Kleinheins, the Head at the City Planning Office. The interview at the Municipality of

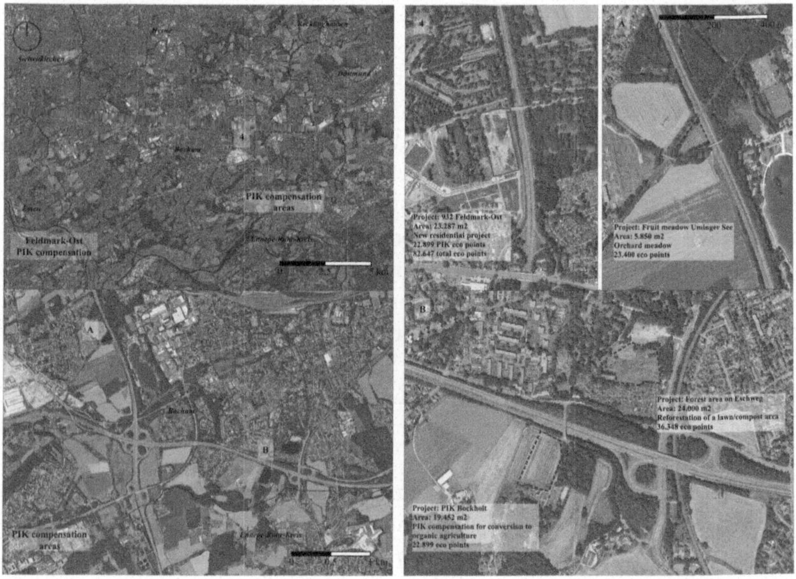

Fig. 7.1 Compensation projects for the new development Feldmark-Ost. (Source: Made by the author with information provided by Stadt Bochum)

Scharpe, one of the project's coordinators, PIK has become a "lighthouse" initiative, fostering collaboration between conservation and agricultural actors, enhancing the municipality's engagement with its farmers, and redefining perceptions of farmland. Rather than being treated as a reserve for inexpensive compensation, agricultural land is now recognized as a valuable tool for ecological enhancement, integrating environmental objectives with productive use.

These cases underscore how local governments can mediate between regional planning goals and on-the-ground farming practices, designing compensation measures that both preserve farmland and support ecological, social, and economic sustainability. They demonstrate a shift in perception: Farmland is no longer merely a space to offset development but an active site for delivering biodiversity, cultural, and educational benefits while maintaining agricultural production.

Bochum was held on July 31, 2024, with Alexandra Scharpe and Clemens Stamm from the Environmental and Green Spaces Office, Lower Nature Conservation Authority.

Land Tenure Regimes and Management Models in the Milanese Countryside

In Milan and Lombardy, public institutions own significant shares of agricultural land, creating opportunities for innovative land management. This section explores the interplay between land tenure and management models, understood as deliberate strategies by landowners to govern their estates responsibly. Two illustrative cases, the Municipality of Milan and Fondazione Patrimonio Ca' Granda, demonstrate how institutional landownership enables experimentation with tenancy contracts, territorial strategies, and even supply chain development.

According to 2020 ISTAT data, of Lombardy's 45,210 farm holdings, 13.2% rely entirely on leased land, while 43.7% combine full or partial leasing with ownership (Table 7.3). In terms of Utilized Agricultural Area

Table 7.3 Tenure status of holdings and UAA in Lombardy 2020 (Unit of measure: number of holdings and hectares)

Type of tenure	*Farm holdings*	*%*	*Total Italy*	*Utilized agricultural area—UAA*	*%*	*Total Italy*
Only ownership	18,729	41.4%	*58.6%*	210,669	20.9%	*33.3%*
Only leasing	5980	13.2%	*10.1%*	158,600	15.7%	*18.6%*
Use rights free of charge	3566	7.9%	*6.0%*	40,679	4.0%	*3.9%*
Ownership and leasing	9180	20.3%	*12.5%*	352,446	35.0%	*27.4%*
Ownership and use free of charge	3166	7.0%	*8.7%*	33,921	3.4%	*5.9%*
Leasing and use free of charge	2342	5.2%	*1.3%*	111,712	11.1%	*3.7%*
Ownership, leasing, and use free of charge	2247	5.0%	*2.7%*	98,959	9.8%	*7.2%*
Total Lombardy	**45,210**			**1,006,986**		
Total Italy			**1,133,023**			**12,535,357**

Source: Elaborated by the author with ISTAT data

(UAA), farms leasing[2] their entire holding represent 15.7% (158,600 hectares) of the region's total 1,006,986 hectares, while those combining ownership and leasing account for 71.7% (721,717 hectares). Compared to the national average, Lombardy has fewer farms leasing their entire holdings (13.2% and 10.1% nationally) but a higher incidence of mixed-tenure arrangements (20.3% and 12.5% nationally). Interviews revealed that many farmers actively manage mixed-tenure holdings, often leasing additional land to supplement family farm operations. For example, Cascina Femegro partially leases neighboring land to extend production, while Cascina Battivacco leases extra land, even in other provinces, to accommodate new business activities.

Regarding tenure and management, 79.9% of holdings are registered as professional entrepreneurs or sole proprietors, lower than the national average of 93.5%. Partnerships represent 17.1% and corporations 2.1%. Among 46,780 farms in Lombardy, 77.6% are managed by men and 22.4% by women, with female managers controlling only 14.5% of total UAA. Most female-managed farms are small, under 5 hectares (61.3%), and predominantly managed by older farmers, with 82% over 45 years old. Male-managed farms are roughly triple in number, but age distributions are similar. Across Europe, 57.6% of farmers are over 55 years old, a trend mirrored in Lombardy and Italy, where 67.3% exceed this age threshold.

Historical trends indicate a clear shift toward leasing. Between 2010 and 2020, farms leasing their entire surface increased by 49.7%, while those fully owned and farmed decreased by 44.1%. UAA under full lease rose by 71.2%, while fully owned and farmed UAA dropped by 28.3%. The most significant growth occurred in farms partially leasing or partially granting land for free use, which increased by 116.2% in holdings and 197.5% in UAA. These figures reflect the pressures of generational transition, as retiring farmers adjust land management and create opportunities for newcomers. As observed in Barcelona, generational succession is a critical challenge, but institutions can actively intervene through policies and flexible mechanisms, such as land banks or programs pairing older farmers with new entrants. The following cases will illustrate how the Municipality of Milan and Fondazione Patrimonio Ca' Granda actively

[2] This number adds all the values on Table 7.3 that includes 'leasing' within their types of tenure. Which are the following categories: Only leasing; Ownership and leasing; Leasing and use free of charge; Ownership, leasing and use free of charge.

manage farmland, integrating these practices into broader governance and planning strategies.

Municipality of Milan

The Municipality of Milan owns approximately 800 hectares of agricultural land, a figure that has grown over the past two decades mainly through expropriations and new development compensation rights following the building boom and speculation crisis of the 1980s and 1990s. Total agricultural land (SAU) in Milan increased from 2910 hectares[3] in 2010 to 5548 hectares in 2020, meaning the Municipality owns roughly 15% of the city's farmland. Within Milan's administrative boundaries, there are 117 active farms, 91 of which are primarily registered in Milan. According to municipal data, 77.7% of the UAA is arable land, 13.4% permanent grasslands, and 8.9% permanent crops, with two hectares dedicated to kitchen gardens (*orti*).

The city-owned farmland is managed by the Agricultural Section, part of Milan's Food Policy Area—a legacy of Expo 2015 and the Milan Food Pact—which oversees the city's Food Policy. This organizational choice reflects a strategic shift in 2007 toward recognizing the potential of Milan's agricultural areas, initially as public spaces and later as food-producing assets. A key early measure was restructuring leasing contracts, extending durations from 1 year to 15–30 years and embedding social and cultural obligations for tenants, encouraging multifunctional activities integrated with farming.

In Milan, the profile of tenants on institutional farmland has evolved beyond conventional farmers to include associations and organizations with social objectives. Farmland contracts can be broadly classified into three categories. The first encompasses agricultural fields without built infrastructure, leased primarily for cultivation. The second includes historic farm buildings, or *cascine*, which are no longer directly linked to farmland and are often repurposed for educational, social, or cultural activities. The third category covers fully functioning farms that combine land and buildings, and are eligible for long-term contracts of up to 30

[3] Data provided by the Municipality of Milan based on the 2010 agricultural census and considered by the 2019 Piano di Governo del Territorio – PGT, the city's main governance and planning instrument, which anticipated to add an additional 322 hectares of SAU.

years, allowing tenants to invest in sustainable production and multifunctional activities.

Fifteen tenants currently operate multifunctional farms under this system, maintaining historic buildings is a key factor in management decisions. Past practices often shifted farmhouses toward non-farming uses (Agostini et al., 2006), particularly where urban encroachment occurred (e.g., Cascina Cuccagna). The new contracts aim to balance structural upkeep responsibilities with tenants' financial capacity, including optional partial leasing of farm buildings, though this has not yet been implemented. Rent is set in line with market prices—€29 per milanese perch (*pertica milanese*)[4] for 2024/2025, with lower rates for non-irrigated land—generating roughly €300,000 annually. Previously, under the State Property and Heritage Unit, farmland was considered a burden, however, now, the Agricultural Unit has established a dual strategy: reducing ordinary expenses while encouraging tenants to adopt environmental and social practices. Future contracts will favor proposals that implement less environmentally harmful farming methods. While Milan's traditional production (rice and fodder) limits direct intervention, the city facilitates diversification and transitions toward horticulture without assuming economic risk.

Two notable strategies support this approach. First, the public-school canteen procurement system creates a guaranteed market for farmers. Second, the creation of the Distretto Agricolo Milanese (DAM) fosters cooperation among farms. Established under the 2009 regional law (D.g.r.n.8/10085), the DAM consortium initially brought together over 30 farms totaling 1155 hectares (40% of Milan's 2010 UAA) to jointly supply food, engage in landscape-environmental revitalization, preserve architectural heritage, and strengthen ties with local communities. Although collaboration has waned, the consortium influenced farm activities and food supply to public schools.

More recent initiatives include farmers' markets in public spaces (*mercati contadini*), now a flagship project of Milan's Food Policy. These markets are open to farmers within Milan and surrounding municipalities, though most participants are from outside the city due to commitments to direct sales. Another program involves farmers in landscape maintenance across urban agricultural parks (Ticinello and Vettabbia) and larger

[4] A measurement unit used for measuring agricultural land. 1 milanese perch (*pertica*) equals to 6,545,179 sqm. 1 hectare equals to 152,784 milanese perches.

metropolitan parks (Lambro, Forlanini, and delle Cave), sometimes through cooperative arrangements, providing additional income.

Over the past 15 years, Milan has developed a management model for its agricultural land that leverages milestones such as Expo 2015, international exposure via the Milan Food Pact, consolidation of the Food Policy Area, and growing citizen interest in local food systems. However, the governance crisis of Parco Agricolo Sud has led the Municipality to focus on a narrower, city-centered model. Despite this, Milan is consolidating a framework that recognizes the strategic potential of public land, integrating social, environmental, and economic objectives within its administrative boundaries.

Fondazione Patrimonio Ca' Granda

Landownership has repeatedly emerged in this research as a crucial factor shaping agricultural production, largely through the conditions imposed on tenant farmers. In Lombardy, large estates were historically owned by religious institutions, noble families, and entrepreneurs—a legacy that echoes the Catholic Church's and elite families' role in food production and local agrarian economies for various centuries. With industrialization, many noble families shifted toward entrepreneurship, while farming gradually lost its place in elite social structures. Peasant families from Brescia, Bergamo, and Lodi migrated to Milan, becoming long-term tenants of these estates.

Fondazione Patrimonio Ca' Granda exemplifies this historical trajectory while introducing a modern, active management model. The Foundation currently owns and manages approximately 8500 hectares between the Ticino and Adda rivers (see Fig. 7.2), of which 6500 hectares fall within protected areas, including Parco Agricolo Sud, Parco Lombardo della Valle del Ticino, and Parco Adda Sud. The estate includes over 130 farmhouses (*cascine*) and more than 200 km of irrigation canals (*rogge*). Since 2014, the Foundation has emphasized its mission as a *heritage steward*, valuing not only the built heritage, but especially soil and water as critical public assets essential for future food production.

The Foundation was established to manage assets donated to Ospedale Maggiore di Milano, founded in 1456 by Francesco Sforza, Duke of Milan. Health system reforms in the 1970s neglected external assets, leading to a deterioration of land and infrastructure. By 2014, the non-profit

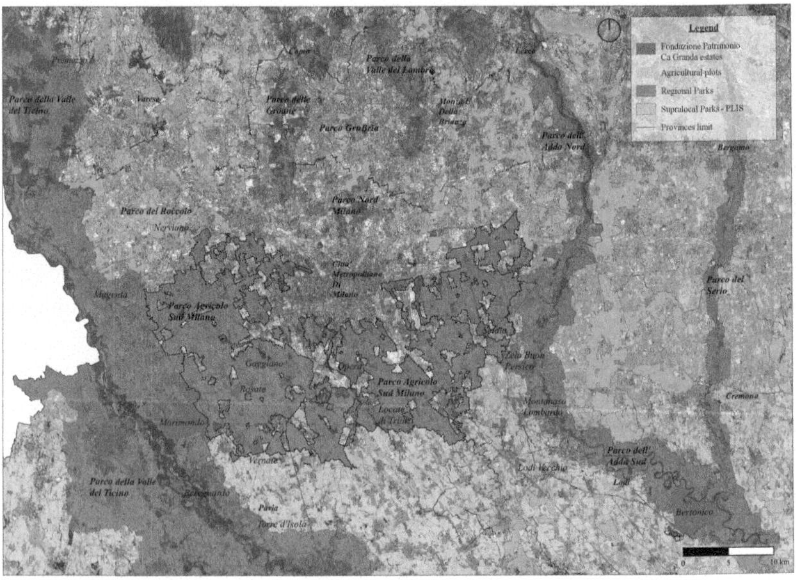

Fig. 7.2 Farming estates owned by Fondazione Patrimonio Ca' Granda. (Source: Elaborated by the author with information provided by Fondazione Patrimonio Ca' Granda)

Fondazione Patrimonio Ca' Granda[5] was created to manage these properties and support scientific research at Policlinico di Milano. Early actions included restructuring tenancy contracts with standardized rents and clear responsibilities for farmers, often replacing inactive tenants to protect farmland, irrigation systems, and built heritage. Current contracts reflect collective agreements with key farmers' associations.

This active management model positions Ca' Granda as a liaison between farmers, the agri-food sector, and regional authorities, enabling diverse initiatives. Notable examples include the 2017 launch of a branded line of milk, yogurt, and rice in partnership with tenant farmers and Oasi Ca' Granda, a program of over 200 activities encouraging local residents

[5] Initially called Fondazione Sviluppo Ca' Granda, the foundation has 21 staff members and is supported by a scientific board of experts. Its main activities include identifying funds to support these programs. In 2023, it raised €2.1 million.

to engage with these productive territories through both farming and nature-based initiatives.

Ca' Granda prioritizes sustainability, acknowledging both the European Union's regulatory pressures and the essential role of farmers in achieving a gradual ecological transition. As explained by a senior official of the Foundation, maintaining soil quality and safeguarding water resources are central to its mission, enabling the monitoring of land use and tenant farming practices. Their approach to sustainability emphasizes collective responsibility, understanding that the supply chain depends on all actors adopting environmentally conscious practices. The guiding principle, "*Abbiate cura di ciò che vi è stato donato*,[6]" underpins projects that monitors its assets and develop mitigation strategies in collaboration with farmers.

Some of Ca' Granda's main projects include monitoring water, soil, and biodiversity through a network of hundreds of weather and hydrological monitoring stations distributed along 200 km of waterways, developed in response to the 2022 water crisis and to anticipate the impacts of climate change. Ecological restoration initiatives have resulted in the planting of 18,000 trees and shrubs, the creation of 30,000 m^2 of new wetlands, and the establishment of 100,000 m^2 of permanent grasslands and flooded meadows (*marcite*), benefiting a range of bird, amphibian, and insect species. Soil quality surveys across 1200 fields, conducted in-house, indicate an average organic matter of 3%[7] and CO_2 sequestration equivalent to 1.5 million tons—comparable to a forest of 850,000 trees 25 m tall.[8] While the Foundation cannot dictate farming practices, these scientific insights serve as evidence-based incentives for tenant farmers to adopt sustainable methods, fostering both environmental stewardship and informed decision-making

To further support farmers, Ca' Granda launched Accademia Ca' Granda, offering free, tailored courses in partnership with national and regional agriculture associations and the philanthropic foundation

[6] A motto adopted in 2019, which was a phrase written in an old clock hung in one of Ospedale Maggiore's buildings. The motto can be translated to 'guard what has been entrusted to you' as a call for stewardship

[7] According to Cornell University's Department of Crop and Soil Sciences, productive agricultural soils have between 3 and 6% organic matter. Data gathered from the Agronomy Fact Sheet Series – Fact Sheet 41.

[8] This information was gathered from Fondazione Patrimonio Ca' Granda's 2023 Social Report (Fondazione Patrimonio Ca' Granda, 2024).

Fondazione Cariplo. The program, held from January to February 2024, focused on the region's primary productions: milk and rice. Seminars combine expert presentations with practical insights, followed by discussions and networking. Topics cover sustainability, EU regulations (CAP), and animal protection, fostering knowledge exchange and reflection on transitioning toward environmentally responsible farming. Farmer feedback has been positive, with a 92% approval rating in the 2023 Social Report. The Accademia illustrates the Foundation's role in capacity building, bridging academic knowledge, regulatory frameworks, and practical farming challenges. It complements Ca' Granda's broader initiatives, from tenancy reforms to biodiversity projects, demonstrating how institutional landowners can actively support farmers in a collective socio-ecological transition.

The Milanese cases of Fondazione Patrimonio Ca' Granda and, to a certain extent, the Comune di Milano highlight the potential of large institutional landowners to advance agricultural innovation, sustainability, and community engagement. Their strategies—including tailored contracts, communication, technical support, monitoring, and education—equip farmers with the resources needed to navigate ecological and economic transitions. At the same time, these cases underscore the challenges posed by fragmented governance, where agricultural practices, political priorities, and business models often come into conflict. Through these cases, it becomes evident that the future of Parco Agricolo Sud will ultimately depend on bridging these divides through coordinated governance, collective action, and the long-term stewardship of agrarian territories.

This chapter focused on a theme that proves highly relevant and crucial for the socio-ecological transition of territories. While the challenges differ across contexts, a consistent finding is the central importance of land access and security, which provides farms with the stability needed to transition to more sustainable farming methods. In Barcelona, land insecurity is closely linked to the difficulty of sustaining a historically rooted farming model, which, despite its strong cultural ties and identity, appears increasingly unsustainable. Addressing this requires not only new mechanisms and initiatives to prevent farmland abandonment but also a cultural shift in the definition of *farmer*, enabling the inclusion of new profiles across age, gender, and racial lines.

In the Ruhr, the main challenge is the rise of industrial agriculture, reflected in the increasing average farm size and the declining number of holdings. Although this trend is common across Europe, agricultural production in the region remains only partially oriented toward food provision. Institutional landowners, such as RVR and the Bochum Municipality, play a potentially pivotal role through strategies that support farmers transitioning toward diversified production and meeting local market needs. Milan presents a similar dynamic, with large institutional landowners shaping the peri-urban agricultural landscape. The case of Fondazione Patrimonio Ca' Granda illustrates that these actors can be central to fostering sustainable and multifunctional agriculture, provided that their commitment is strong and their strategies balance the needs of farmers with long-term environmental objectives.

References

Agostini, S., Pizzingrilli, P., & Rausa, P. (2006). *Il patrimonio rurale vernacole ai margini della metropoli. Analisi di quattro insediamenti nell'area metropolitana milanese.* Libreria CLUP.

Baysse-Lainé, A., & Perrin, C. (2018). How can alternative farmland management styles favour local food supply? A case study in the Larzac (France). *Land Use Policy, 75*, 746–756. https://doi.org/10.1016/j.landusepol.2018.03.012

Beingessner, N., Magnan, A., & Wendimu, M. (2023). "Land imaginaries" in Western Canada: (Financial) neoliberalism, agrarianism, and the contemporary politics of agricultural land. *Environment and Planning C: Politics and Space, 41*(4), 637–655. https://doi.org/10.1177/23996544221144168

Bottmeyer, M. (2011). Land Management of Former Industrial Landscapes in The Economic Metropolis Ruhr. *Bridging the Gap Between Cultures.* FIG Working Week.

Calo, A., McKee, A., Perrin, C., Gasselin, P., McGreevy, S., Sippel, S. R., Desmarais, A. A., Shields, K., Baysse-Lainé, A., Magnan, A., Beingessner, N., & Kobayashi, M. (2021). Achieving food system resilience requires challenging dominant land property regimes. *Frontiers in Sustainable Food Systems, 5, 683544.* https://doi.org/10.3389/fsufs.2021.683544

Christenn, K. J., & Timpe, A. (2019). *Postersammlung Co-Produktionsprojekte im Grünzug Östliches Emschertal: CoProGrün Projektplattform am 30.04.2019 in der DeusenKirche in Dortmund* [Online]. CoProGrün Projektplattform in der DeusenKirche in Dortmund.

European Commission. (2023). *Contracts2.0 Project. Co-design of novel contract models for innovative agri-environmental-climate measures and for valorisation of environmental public goods.* CORDIS. https://doi.org/10.3030/818190

Fondazione Patrimonio Ca' Granda. (2024). Bilancio sociale 2023. https://www.fondazionepatrimoniocagranda.it/wp-content/uploads/2024/06/bilancio-sociale-fondazionepatrimonio-2023.pdf

Otsuka, K., Chuma, H., & Hayami, Y. (1992). Land and labor contracts in Agrarian economies: Theories and facts. *Journal of Economic Literature, 30*(4), 1965–2018.

Regionalverband Ruhr. (2023). *Landwirtschaftliche Leitlinien.* Regionalverband Ruhr.

Rotz, S., Fraser, E. D. G., & Martin, R. C. (2019). Situating tenure, capital and finance in farmland relations: Implications for stewardship and agroecological health in Ontario, Canada. *The Journal of Peasant Studies, 46*(1), 142–164. https://doi.org/10.1080/03066150.2017.1351953

Sereni, E. (1961). *Storia del paesaggio agrario italiano* (1st 2020). Editori Laterza.

Sevilla Buitrago, Á. (2022). *Against the commons. A radical history of urban planning.* University of Minnesota Press.

CHAPTER 8

Food Networks and New Urban–Rural Relationships

Introduction

The rise of alternative food networks has been one of the most significant developments in food production, distribution, and consumption at the turn of the century (Goodman et al., 2013). This trend is closely linked to the resurgence of urban agriculture and has challenged the dominance of conventional, industrialized "corporate food regimes" (Holt Giménez & Shattuck, 2011) by promoting shorter, more localized supply chains between producers and consumers.

One prominent model is Community-Supported Agriculture (CSA), in which farmers enter formal agreements with a fixed group of consumers, creating a system based on solidarity, shared risk, and mutual support (Enthoven & Van den Broeck, 2021). This approach has allowed farmers to gain financial stability and, in many cases, to transition from conventional or specialized production toward ecologically certified and sustainable practices. Interviews conducted during this research (discussed in the previous chapters) confirmed that CSA and similar networks often include a strong community and social component, enhancing cohesion and mutual engagement.

However, the corporate food system has not remained static. Multi-stakeholder conglomerates and industrial producers have adapted to consumer demand for *healthier* and more environmentally conscious products, a process described by Friedmann (2005) as "green capitalism." Despite

© The Author(s), under exclusive license to Springer Nature Switzerland AG 2026

A. J. Salvador, *Metropolitan Food Systems*, Sustainable Urban Futures, https://doi.org/10.1007/978-3-032-26016-1_8

these shifts, access to healthier, sustainably produced food remains uneven and frequently a social privilege (Born & Purcell, 2006). As Brunori (2007) emphasizes, the narrative of place and producer, the local embeddedness of products, and effective communication strategies are central to building a cohesive and informed local food system. These dimensions allow consumers to connect more directly with farmers and their practices, fostering awareness and participation in sustainable food networks.

From the farmers' perspective, these networks are not merely commercial arrangements. They provide an opportunity to rethink business models, diversify production, and engage in socially and environmentally meaningful practices, exemplified with the cases in Chap. 6. Moreover, many of these initiatives are self-organized, reflecting broader forms of community cohesion and collective agency. Others emerge through policy support, facilitating closer interactions between producers and consumers while circumventing industrial supply chains. This chapter highlights a variety of examples collected through field visits, interviews, and secondary sources, illustrating the multiple ways in which alternative food networks contribute to sustainable food production, local engagement, and social cohesion.

Governance and Management of Peri-urban Agrarian Systems in the Province of Barcelona

The Baix Llobregat Agricultural Park consortium functions as the governance and management body of the park, with responsibilities defined by the Management and Development Plan and the implementation of the Strategic Land Use Plan. Since its establishment in 1998, it has brought together the 14 municipalities within the park's boundaries, the Unió de Pagesos as farmers' representative, and multiple levels of government, including the Diputació de Barcelona, the Metropolitan Area of Barcelona (AMB), the Baix Llobregat County Council, and the Government of Catalonia through its Department of Climate Action, Food, and Rural Agenda. These actors were involved from the park's early design stages. The consortium's governance structure is organized into five bodies: the Plenary Council, the Presidency and Vice-presidency, the Executive Commission, the Management Entity, and the Agrarian Council.

The management entity operates with a dedicated administrative team of approximately 10 staff members and is physically located within the park, in the refurbished historic farmhouse of Can Comas. This on-site presence is strategic, facilitating daily interaction with farmers and close engagement with ongoing agricultural activities. The park manager represents the institution within the consortium and coordinates a wide range of programs beyond statutory planning duties. These include capacity-building initiatives, technical assistance, and, increasingly, projects supporting environmentally oriented farming practices. As noted by the park's administrative staff, this shift has generated tensions with conventional farmers, particularly as the park has begun to actively support an emerging generation of ecological producers. Can Comas also hosts the offices of the *Agrupacions de Defensa Vegetal – ADV*[1] (plant protection associations), which assist farmers in adopting more sustainable practices, especially in reducing the use of pesticides.

Promoting and Building a Territorial Brand

As outlined in the Baix Llobregat Agricultural Park's Management and Development Plan, communication and marketing are central to the economic viability of local farms. The core branding strategy is the *Producte fresc* label, which identifies products originating from the park and is widely visible in farmers' markets and local grocery stores. Complementary initiatives such as *Sabors de l'Horta* connect producers with local restaurants, using gastronomy as both a marketing and educational tool that highlights seasonal products and farming cycles.

The park's management entity also supports food festivals centered on geographically recognized products—such as El Prat artichokes, Gavà asparagus, and Santa Coloma de Cervelló cherries—actively promoted through social media and local press. Digital platforms further reinforce this strategy by publishing farm profiles and maps of on-farm shops (*agro-botigues*). In recent years, communication efforts have increasingly highlighted ecological producers as a way to incentivize broader transitions toward environmentally oriented practices. However, farmers' markets (*mercats de pagès*) represent the most significant distribution and marketing tool. Beyond their economic relevance, these weekly markets—typically located 2–8 km from the farms and set in prominent public

[1] ADV are collective farmer organizations focused on integrated plant protection, crop health monitoring, and the reduction of chemical inputs through coordinated action and technical support.

spaces—enable direct interaction between producers and consumers. Collectively, these initiatives contribute to positioning peri-urban agriculture and the Baix Llobregat farming culture within Barcelona's urban imaginary.

At the municipal level, El Prat de Llobregat has developed its own branding strategy around the *carxofa Prat*[2]. While supported by the park's management entity, some farmers perceive this approach as relatively individualistic. A key actor is the Cooperativa Agrícola de El Prat, which brings together 13 farmers cultivating approximately 250 hectares. Cooperative exemplifies the role of collective organization in ensuring product quality, sharing resources and knowledge, and jointly negotiating market access. Its products are sold through on-farm shops, the weekly farmers' market, and a dedicated pavilion at the Mercabarna wholesale market. This collective model demonstrates how branding and communication can reinforce cooperation and strengthen the economic and cultural sustainability of peri-urban agriculture.

The role of the management entity thus combines everyday technical work with a strong political and mediating function, requiring continuous coordination with municipalities, higher levels of government, farmers' organizations, cooperatives, and consumers.

An Agrarian Space as a Unit of a Larger System

The concept of *espacio agrario* (Spanish) or *espai agrari* (Catalan) appears frequently in local planning documents, but its meaning goes beyond simply denoting areas of agricultural activity. The planning of the Baix Llobregat Agricultural Park and the broader Barcelona food system reveals that an *agrarian space* embodies decades-long processes, including collective efforts by local peasants and farmers (*pagesos*) from the 1970s and 1980s across Catalonia. Regions such as El Maresme and the Delta de l'Ebre witnessed initiatives inspired by agronomist Josep Montasell (Paül, 2015). While many failed to materialize, they remain significant examples of collective action, illustrating the evolution of agrarian concepts. Montasell's contributions were pivotal in defining the unique

[2] The initiatives are jointly led by the Municipality of El Prat and the *Cooperativa Agrícola d'El Prat*, in partnership with restaurants and local retailers and promoted through public events. While the artichoke was previously branded as a product of the agricultural park, current efforts focus on obtaining PGI status for the *carxofa del Prat*.

characteristics of agrarian spaces, influencing planning proposals such as the 1996 vision for 11 agricultural parks.

Montasell's process culminated in the establishment of the Baix Llobregat Agricultural Park, transforming the abstract notion of agrarian space into a planned and managed territory. Montasell as the Baix Llobregat park's first director, also contributed to the development of the legislative framework of the 2010 *Carta de la Agricultura Periurbana* (Peri-urban Agriculture Charter), which later informed the regional law *Ley 3/2019 de los Espacios Agrarios* (Law 3/2019 on Agrarian Spaces). These documents embody decades of activism, lobbying, and management, supported by stakeholders such as the Unió de Pagesos, cooperatives, and multi-level government agreements.

Law 3/2019 elevates peri-urban agriculture within the political agenda, defining it as activity "carried out in peripheral urban areas" with environmental, cultural, food, and forestry dimensions. It emphasizes professional farming, excluding nonprofessional agriculture, and distinguishes an *agrarian space* from a generic *agricultural space* by explicitly including livestock and forestry. The law provides guidance on planning and management, delegating responsibilities across government levels and highlighting the coordination required among municipalities spanning multiple jurisdictions.

The 2010 Peri-urban Agriculture Charter frames agrarian spaces through preservation, planning, development, and management. Drawing from the Baix Llobregat experience, it outlines strategic actions across governance, territorial protection, productive activities, commercialization, landscape, cultural heritage, and biodiversity. An agrarian space thus emerges as a distinct unit—productive, social, environmental, and spatial—serving as a landscape, a production entity, and a recognizable territory with a unique identity (Salvador, 2024). It is scalable but transcends specific interventions, bridging open-space management, green systems, and food production while encompassing cultural, social, economic, and environmental dimensions.

Although the full system of 11 parks was never realized, it shaped the approach to recognizing these areas beyond generic farmland and inspired a current strategy of new agricultural parks by Barcelona's provincial government (see Fig. 8.1). This recent strategy is coordinating and supporting local consortiums for the creation and recognition of additional agrarian spaces, each defined by spatial, productive, and social

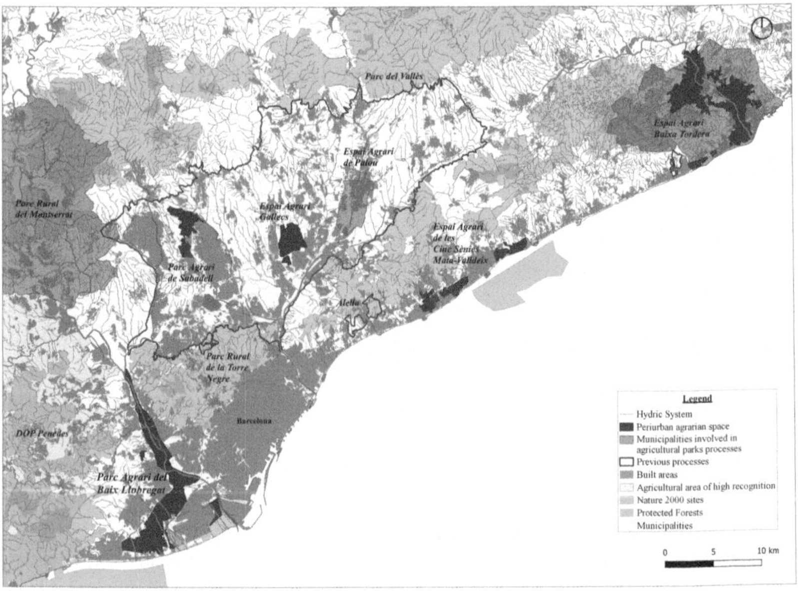

Fig. 8.1 Agrarian spaces of the Barcelona metropolitan area. (Source: Elaborated by the author with data and information from various sources)

characteristics. Currently, there are nine agricultural parks totaling roughly 109,000 hectares.

Studying these spaces in detail clarifies the notion of *local* from a production perspective, revealing their resources, challenges, and socioeconomic structures. While agricultural parks exemplify planned spaces, alternative models, such as *huertas* or *vegas*, highlight historical and self-organized production areas, including cooperatives and public lands, demonstrating the diversity of agrarian configurations.

Espai Agrari Baixa Tordera

This peri-urban agricultural area has over 200 years of food production history across five municipalities. Local farmers' cooperatives collaboratively established a management entity, a process beginning in the 1990s, gaining formal momentum in 2006 with a municipal agreement, and materializing in 2017. In 2020, Barcelona's provincial government commissioned the Management and Development Plan, aligning with the 2010 Peri-urban Agriculture Charter framework.

The Espai Agrari Baixa Tordera's 2020 Management and Development Plan emphasizes strategies for transitioning to an agroecological model. Farms, averaging 4–6 hectares, primarily produce vegetables and fruits for local consumption. The attractive landscape, enhanced by proximity to the sea, is valued by citizens but threatened by urban encroachment. The Tordera River connects the area to ecologically significant sites, forming a region of "biographic richness" (Diputacio Barcelona, 2020). While agricultural boundaries are not precisely defined, management encompasses all five municipalities, potentially exposing land to urban expansion. Agriculture coexists with forestry and other land uses, covering 60% of the area, though biodiversity and wildlife can pose production challenges. The plan addresses the region's territorial complexity, avoiding a purely sectorial approach and embracing the broader agrarian concept outlined in the peri-urban agriculture charter and law.

Coexistence with Natural Systems and the Transition to Ecological Farming

The agricultural park occupies a strategic location near Barcelona, its satellite towns, and metropolitan infrastructure. Yet this area is also ecologically significant given it hosts diverse avian species in wetlands within and adjacent to the park and along the Llobregat Delta. Over the past decade, environmental groups have advocated expanding the Special Protection Area for Birds (SPA or ZEPA in Spanish) into a larger portion of the park as compensation for the airport's 2009 expansion. This area forms part of the broader Baix Llobregat-Garraf Marine Area and the adjacent SPA Costas del Garraf. Farmers strongly oppose this expansion, citing potential disruptions to agricultural practices and production. A third perspective favors further airport expansion, directly affecting the Llobregat Delta's protected areas. These conflicting visions reflect broader governance challenges: Biodiversity and agriculture are often managed separately despite being part of the same *green system*. The debate has become polarized and politicized, sidelining the pressures of industry and infrastructure expansion.

Under El Prat's jurisdiction, some farming already follows SPA guidelines, but further expansion faces opposition, especially in areas valuable for agriculture and water access. SPA restrictions would prohibit greenhouses, reduce pesticide use, and require coordinated production calendars compatible with Delta fauna. The Unió de Pagesos actively lobbies against these measures highlighting potential detrimental economic impacts.

The agricultural park's management entity plays a key role in supporting farmers, providing guidance on sustainable practices. For example, promoting integrated pest management and agroecological techniques through the *Agrupaciones de Defensa Vegetal* (ADV). A few pioneering producers have adopted these methods, becoming advocates for a broader transition beyond merely local production. The Network of Municipalities for Agroecology (*Red de Municipios por la Agroecología*), which includes El Prat de Llobregat as the sole member among the park's 14 municipalities, further strengthens this shift.

Despite these efforts, transitioning to ecological farming methods remains limited. The 2020 Spanish Agrarian Census reports ecological farming represents only 9% of Baix Llobregat, comparable to El Maresme (8%), while Penedès, a wine-focused production area, reaches 40%. The park's management entity has not driven a unified transition but has supported complementary initiatives, such as marketing ecological farms, providing certification guidance, and hosting educational programs. Institutions such as the Mercabarna wholesale market have also opened dedicated spaces for ecological products, including a Biomarket, although its use remains limited. During a visit, about half of 22 stands were occupied. An interview held with a representative of Mercabarna suggests that administrative demands and financial investments continue to slow farms transitioning to ecological farming methods.

As evidenced through previous chapters, in this context, family farms often undertake multiple responsibilities beyond production, including food processing, sales, logistics, client engagement, and accounting. Eco Central, founded by Michelin-starred chef Alejandro Guzmán in 2007, exemplifies innovative solutions to these challenges. Initially conceived as a food basket delivery service, Eco Central has evolved into a multifaceted operation. It now runs a store in Barcelona's Gràcia neighborhood and, most notably, functions as a specialized catering supplier, serving over 80 school canteens and approximately 17,000 meals daily. They collect products directly from farms, coordinate logistics, support farmers with planning their farming calendars, and support schools transitioning to eco-friendly canteens. Located in L'Hospitalet de Llobregat, about 5 km from the agricultural park, Eco Central works with over 40 Catalan farmers, with a particular emphasis on supporting female farmers and fostering collaboration and knowledge exchange among producers. Farmers' stories are showcased both in the shop and online, highlighting the social and environmental value of their work. Through initiatives like Eco Central, the transition toward agroecological production is facilitated, integrating ecological farming with practical market access and education while strengthening the social fabric of local agriculture.

The Barcelona case clearly benefits from its specialization in fruit and vegetable production and its proximity to key infrastructure, which enables direct sales and the structuring of a functional km-0 food system. This condition has been further reinforced by the agency and long-term commitment of institutional actors, particularly Barcelona's provincial government and the quarter-century of work carried out by the Baix Llobregat Agricultural Park management entity. At the same time, the case reveals a system that is actively engaging with other production territories and highlights the importance of acknowledging the specificity of these individual units (*espacios agrarios*), both in productive and social terms.

The mechanisms used to govern agrarian spaces have evolved, as exemplified by the Baixa Tordera case, alongside shifting objectives, particularly the promotion and support of transitions toward ecological farming methods. The Baixa Tordera example also illustrates how planning and land-use instruments, which were foundational to the Baix Llobregat Agricultural Park, are increasingly being replaced by policies and development strategies, an evolution that calls for ongoing monitoring and critical reflection. Finally, this case underscores the agency of diverse actors, of committed institutions and external groups alike, working to advance innovative farming models and food networks, such as the Cooperativa Agrícola de El Prat and Eco Central.

The Role of Planning and Design in Shaping Food Networks in the Ruhr Metropolis

As discussed in earlier chapters, a central challenge in the Ruhr Metropolis lies in steering agricultural activity more clearly toward food production while engaging with bottom-up practices that have long existed on the urban fringe, such as allotment gardens (*Schrebergärten* or *Kleingärten*) developing alongside commercial agriculture (Fox-Kämper et al., 2023). Farms are increasingly responding to this challenge, and initiatives such as the CoProGrün program led by Regionalverband Ruhr—discussed in earlier chapters—illustrate how local governments, and landowners, are beginning to support this transition. Access to traditional allotment gardens, however, remains highly restricted. In response, newer forms of community food production have emerged, some directly organized by farmers, including rent-a-field schemes that offer alternatives for groups often excluded from *Kleingärten* structures, particularly younger generations and newcomers. These productive spaces are especially relevant in the Ruhr context, as they challenge conventional notions of public space and parks, engage with environmental reclamation processes, and generate forms of social interaction centered on food production and leisure.

At the same time, these practices and their cultural perception have been at the center of debate within one of the Ruhr Metropolis's most significant initiatives: the KuLaRuhr project: *Sustainable Urban Cultural Landscape in the Ruhr Metropolis* (2011–2014). The project focused on sustainable management of land, water, and energy within the Emscher Landscape Park and made a key contribution by clearly distinguishing between *urban agriculture* and *urban gardening*. Urban gardening was framed as a largely hobby-based activity with strong social and cultural value but limited spatial impact, while urban agriculture was defined as professional, market-oriented production with significant spatial implications, occurring mainly in peri-urban areas characterized by high land values. This form of agriculture is associated with diverse crops and multifunctional business models oriented toward urban markets. KuLaRuhr also highlighted the limited visibility of professional urban agriculture compared to urban gardening, prompting strategies focused on communication and increasing farmers' public presence.

The following cases examine hybrid structures that seek to bridge the entrenched dichotomy between urban gardening and urban agriculture by enabling the coexistence of different productive practices. Conceived as public spaces, these parks result from planning and design processes that reintroduce productive dimensions into landscapes traditionally framed as recreational, reflecting historical forms of coexistence between agriculture, public space, and urban society (Timpe, 2017). These examples are discussed in this chapter to examine the role of such elements, and of urban gardening more broadly, within the structure of local food systems.

Revisiting the Park as a Mechanism for Local Food Systems. Hugo Biomass Park and Hassel District Park

These two parks offer a particularly illustrative perspective on the integration of productive activities within post-industrial landscapes. First, they illustrate how productive activities—such as biomass energy and agriculture—can become central to park design, particularly in the reprogramming of former industrial sites like the Hugo colliery and the Hassel coking plant. This highlights the potential of design to guide post-industrial transformation through productivity-oriented strategies (see Fig. 8.2). Second, both sites are owned by RAG Montan Immobilien, one of the Ruhr region's largest landowners, which manages former assets of Ruhrkohle AG, now operating as Evonik Industries AG. The role of large landowners is particularly relevant, as favorable leasing conditions and long-term stewardship can position them as key agents in advancing local food production. Landownership-based strategies thus represent a significant area of innovation as discussed in the previous chapter.

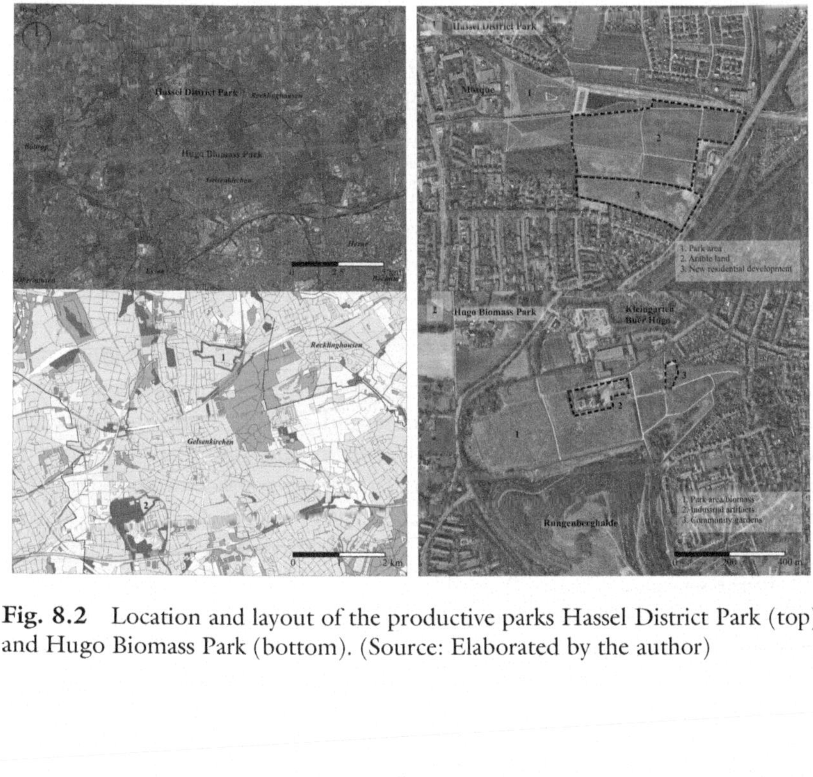

Fig. 8.2 Location and layout of the productive parks Hassel District Park (top) and Hugo Biomass Park (bottom). (Source: Elaborated by the author)

The Hugo Biomass Park occupies 28 hectares of a former colliery and coking plant that operated for over 130 years until 1999, deeply shaping the area's social fabric. The initial proposal by the architectural firm lohrberg stadtlandschaftsarchitektur designated 11 hectares for short-rotation willow and poplar cultivation for energy production. However, this model faced challenges related to drought caused by climate change and logistical constraints, leading to the cessation of biomass production in 2020 when RAG Montan withdrew from the project. Despite this, the park served as an important testing ground for biofuel strategies discussed during the drafting of the 2010 Emscher Landscape Park master plan and influenced parallel projects, including the Hassel District Park (Timpe, 2017). The project was supported by institutions such as the Ministry of the Environment of North Rhine-Westphalia, Ruhr University Bochum, the NRW State Agency for Forest and Wood, and the City of Gelsenkirchen, and was involved in several European and national knowledge-exchange initiatives.

Today, Hugo Biomass Park functions as a public space and an ecological compensation area, comprising 20.5 hectares of landscape and 3 hectares of forest. Additional uses include 1.5 hectares of community gardens (approximately 100 plots) and a dye garden for natural pigments. These are managed by the association *Förderverein Grünlabor im Biomassepark Hugo e.V.*, which collaborates with schools, daycare centers, and civil society groups. The park adjoins historic allotment gardens and retains industrial artifacts and murals, preserving the area's working-class heritage. It also foregrounds the Rungenberghalde, a 115-meter-high tailings heap formed from mining waste. Since its completion in 2015, biodiversity has increased significantly, making the park a site for ecological monitoring, as described during a visit by a senior city official from the City of Gelsenkirchen during a visit in May 2023.

The Hassel District Park covers 36 hectares on a former coking plant and mining site active for 50 years. Conceived as a continuation of the Hugo Biomass Park experience, it emerged from a design competition also won by the architectural firm lohrberg stadtlandschaftsarchitektur. The proposal included a reduced short-rotation coppice area of 10 hectares and incorporated urban park elements such as a skate park and a lake. Due to soil contamination, extensive remediation was required, including the import of approximately 600,000 m^3 of soil. When implementation began in 2017, biomass production was replaced with agricultural use. Opened in 2020, the park allocates 15 hectares to urban agriculture,

3 hectares to flower strips for biodiversity, fruit groves, and over 100 boundary trees. It also includes 5.5 hectares of landscape features, a 2-hectare lake, meadows, grazing areas for sheep, and recreational spaces near residential zones. These uses were defined through participatory design processes. The eastern edge, bordered by roads and railways, prioritizes access for agricultural machinery, while the southern edge adjoins a 6-hectare residential development by RAG Montan Immobilien. The result is a hybrid landscape where agricultural fields function as the primary spatial and aesthetic elements (see Fig. 8.3).

During a site visit in May 2023 and an interview with the farmer leasing the land at the time—who also owns a historic farm in the area—he explained that he had adopted a solidarity-based farming model (*Solidarische Landwirtschaft*—SoLAWi), supported by a consumer partnership. Although he withdrew from farming activities in 2024 due to financial difficulties, the interview provided valuable insight into attempts to integrate alternative agricultural business models within park design

Fig. 8.3 Agricultural fields and paths at the Hassel District Park. (Source: Author)

and management. The farmer, his family, and a young aspiring farmer, described both the opportunities and challenges of this model.

In addition to farming, this family operated a seasonal beer garden and direct-sale shop, offering mainly organic vegetables, though uncertified due to soil conditions and the recent establishment of the activity. Crops were grown on imported soil covering the contaminated brownfield, and products were regularly tested. Despite risks associated with media attention and potential greenwashing—particularly given the involvement of RAG and environmental compensation mechanisms—the farmer emphasized the project's genuine effort to implement circular economy principles. Although the additional 15 hectares exceeded his family's previous farming scale, local reports confirmed his eventual withdrawal from the project.

At first glance, the Hugo Biomass and Hassel District parks may appear as partially unfulfilled experiments. Nonetheless, they are highly relevant to this research, as they demonstrate how governance, planning, design, and farming intersect with the purpose of "re-programming" (Salvador, 2025) agrarian landscapes. Across roughly 30 hectares, these projects reprogram culturally significant yet degraded sites into public spaces with strong social and environmental value. Located in former mining neighborhoods that have become increasingly multi-ethnic, these parks also play an important social role. The Hassel District Park borders a mosque built in 1994, and Hugo's community gardens are actively used by Syrian residents for self-consumption. Given the limited access to traditional German allotment gardens, these newer community-based models—often farmer-led—offer alternative forms of inclusion.

This subchapter illustrates relationships across multiple scales through planning mechanisms and design tools. It contributes to the broader discussion initiated in Chap. 3 on strategies for protecting and revitalizing productive open spaces. The concept of *productive parks* (Timpe, 2019) emerges from a design-based approach situated within the genealogy of parks in Germany and Europe. As Timpe (2019) argues, coordinating diverse land uses within open spaces is essential for sustainability and resilience in polycentric regions, particularly under conditions of urban shrinkage and public budget constraints. In this context, productivity is not merely descriptive but operational, positioning parks as multifunctional landscapes capable of generating economic, social, and environmental value across planning and governance scales, and key resources for inquiring on the role of planning and design in the complex assemblage of local food systems.

Milanese Food Networks: Policymaking as an Emerging Dimension of Food Planning

Expo 2015 marked a turning point for Milan, making it possible to speak of a *before and after* the event. As noted by Susi Botto and Di Vita (2017), the Expo provided Milan with a global platform to position itself as a city of events, urban transformation, and international business and tourism. The lengthy and complex preparatory process also allowed the organizing committee to refine the Expo's thematic focus, shifting from an initial branding-oriented approach to food—rooted in a *Made in Italy* narrative—toward a more critical engagement with global challenges related to nutrition and food sustainability. The final slogan, *Feeding the Planet, Energy for Life,* addressed issues of food production, consumption, and unequal access to healthy food worldwide. Most relevant for this section, the Expo foregrounded the role of local governments in addressing food-related challenges.

Although world expos are typically structured around national pavilions, Milan succeeded in embedding a local and urban agenda through the Milan Urban Food Policy Pact. This initiative marked the beginning of a "translocal" collaboration (Moragues-Faus & Sonnino, 2019) positioning Milan as a key advocate for local food policies and as a hub for the exchange of best practices among cities. Signed by more than 280 cities across 84 countries, the pact has evolved into a network and a *protocol* aimed at addressing food-related challenges at the urban level. It also reinforced Milan's global-city image and strategic engagement in international city networks (Coppola et al., 2020), earning support from local governments worldwide.

A major milestone in the lead-up to Expo 2015 was the creation of the Milan Food Policy Program. Rooted in Milan's historical relationship with its surrounding agricultural landscape—particularly Parco Agricolo Sud—and framed through the notion of a *rural metropolis,* the policy was conceived as a strategic and holistic document addressing the entire food cycle. It placed particular emphasis on urban phases such as consumption, food loss, and waste, which are more directly influenced by public policy. Launched in 2014 as a joint initiative between the City of Milan and the Cariplo

Foundation, the policy was coordinated by a newly established Food Policy Office and supported by a Metropolitan Food Council to involve municipalities within the Metropolitan City of Milan, as well as civil society, universities, private actors, and other institutions. A monitoring system of 44 indicators, developed in collaboration with FAO, was also implemented.

Ten years after its launch, Milan's Food Policy Office remains active, adapting to emerging challenges such as the COVID-19 pandemic, the wars in Ukraine and Palestine, climate change impacts, and localized social and political conflicts. The influence of the Milan Food Policy and the Food Policy Pact is evident globally and particularly across Italy, where numerous cities have developed food policy initiatives (Mazzocchi & Marino, 2020). These initiatives often frame food security within a resilience-building approach, aiming to strengthen local economies, educate consumers, and promote healthier lifestyles. Cities such as Rome, Matera, Torino, Bergamo, Trento, Pisa, and Livorno have adopted food policies or related strategies in varying formats. Torino and Matera launched the *Atlante del Cibo* digital platform, while Bergamo and Livorno have focused more strongly on education around nutrition and local food production.

A notable feature of these processes is the growing cooperation among local governments, academic institutions, activists, and public officials. The Italian Network for Local Food Policies (*Rete Italiana per le Politiche Locali del Cibo)*, founded in 2018 as an academic initiative and later expanded to include civil society and public officials, exemplifies this collaborative approach. The network seeks to define an Italian agenda to local food policies, framing them as an emerging field with significant potential that requires national-level discussion and multi-actor engagement. It functions as a platform for exchange, coordination, and the development of a shared language. One key debate concerns the shift from *urban* to *local* food policies, emphasizing a territorial and process-oriented approach rather than fixed geographic categories. This perspective also reinforces the use of the term *food policies* to encompass nutritional, cultural, social, environmental, and economic dimensions, avoiding narrower interpretations focused solely on nourishment.

Milan's Urban Food Policy and Its Flagship Projects

Milan's Urban Food Policy is built around five key priorities: ensuring healthy food and water for all, promoting the sustainability of the food system, advancing food education, combating food waste, and supporting scientific agri-food research. These priorities highlight both the educational dimension of food and the challenge of equitable access. The policy addresses inequalities in access, loss, and waste, framing them as urban issues shaped by demographic, socioeconomic, and cultural differences across the city. It also emphasizes the creation of a coherent governance framework, connecting public and private actors, grassroots organizations, non-profits, universities, and enterprises across multiple scales. The flagship projects presented below exemplify this integrated approach, demonstrating how a food policy can operate in practice while navigating the complex networks and challenges of an urban food system.

One of the most significant flagship initiatives is the public procurement system of Milano Ristorazione, a municipal catering company that provides over 85,000 meals daily to more than 400 public schools, daycare centers, and elderly care facilities. The municipality recognizes the central role of school meals in children's nutrition, particularly in a context where 1 in 10 children cannot afford an adequate diet. In this sense, the school meal may represent the most substantial daily meal. In 2016, Milano Ristorazione established an agreement with the *Distretto Agricolo Milanese* (DAM), a rural agricultural district comprising 31 farms and approximately 2000 productive hectares. Through this partnership, the city engages local farmers while addressing nutrition, education, and food system sustainability. Complementary educational activities and communication campaigns target families and school staff, promoting healthy eating habits beyond the canteen.

Despite efforts to prioritize bio-certified and local products, sourcing remains challenging. In 2023, only 26 of the 109 products required by Milano Ristorazione were procured locally[3], defined as within a 70 km radius of Milan. The evaluation system rewards km 0 products and short supply chains—those supplying directly from producers—alongside environmental criteria related to farming methods, transport, packaging, and resource management. Price accounts for only 30% of the final score, reflecting a value-based procurement approach rather than a cost-minimization strategy. One successful outcome of this model is the establishment of a short rice supply chain through DAM, supplying

[3] This statement was made by representatives of Comune di Milano and Milano Ristorazione during the EU City Lab on Land Strategies to Feed the City, organized by the URBACT Secretariat in Milan on October 23–24, 2024.

approximately 240,000 kg of rice annually for school meals, valued at €300,000. The total procurement budget for 2023–2024 reached €34 million across 30 tenders. These calls are publicly available, allowing farmers to anticipate demand. The municipality actively encourages diversification among producers in a region traditionally specialized in rice and dairy. Currently, vegetables for preschool canteens are supplied by Cascina Biblioteca, a social farming initiative operating on municipally owned land. Public food procurement has become a central policy instrument discussed internationally through city networks such as the Milan Urban Food Policy Pact framework.

Another flagship initiative is the Neighborhood Food Waste Reduction Hub (*Hub di Quartiere contro lo Spreco Alimentare*), launched in 2018 to reduce food waste. These hubs collect surplus food from supermarkets, canteens, and markets and redistribute it through non-profit organizations. Starting with a pilot in the Isola neighborhood, the program expanded during and after the COVID-19 lockdown to Lambrate, Gallaratese, Corvetto, and Ortomercato, Milan's wholesale market. The initiative is led by the Municipality of Milan with support from a coalition of philanthropic foundations, social welfare programs, business associations, and academic institutions. Each hub operates through locally tailored arrangements, depending on available spaces, partner organizations, and volunteer networks. According to municipal data, in 2023 the hubs recovered over 615 tons of food, reaching approximately 27,000 people and equivalent to 1.23 million meals.

Across these initiatives, the concept of solidarity emerges as a central dimension of Milan's food policy and of broader debates on local food system transitions. This emphasis aligns with a substantial body of academic literature highlighting cooperation and solidarity as foundational principles for alternative food systems (Federici 2009; Tornaghi & Dehaene, 2020).

Food policymaking in Italy operates along two intertwined dimensions: its outward impact, including the capacity to lead a global pact on food-related issues, and the internal implementation of policies and projects at the city level. A key challenge remains the integration of a supra-local scale that acknowledges food production beyond administrative boundaries, while aligning urban-focused policies addressing consumption patterns and social issues with broader territorial strategies—an aspect that this research underlines.

Production and Consumption Agriculture Networks

The analysis of farming styles and business models discussed in the previous chapters is inherently connected to supply chains and food networks, as the destination of production and the organization of distribution are central to farm economies. This topic is addressed separately due to the wide range of collaborative arrangements through which small and family farms have organized collective efforts to reduce the competitive gap with larger corporate producers. These strategies involved not only pioneering business models—as evidenced by the selected case studies discussed in the Chap. 6—but also the construction of shorter supply chains. These chains generate added value through the role of farmers themselves, the characteristics of products, and the social relationships embedded throughout production and distribution processes. Milan, Lombardy, and Italy more broadly constitute territories where such practices have flourished through bottom-up initiatives and, more recently, through food policy and governance efforts. These initiatives have intersected with movements advocating transitions toward agroecology and organic or biological farming practices (Gentili & Zampetti, 2021).

This process has also introduced innovative forms of cooperation among farmers, namely, agricultural or production districts (*distretti agricoli*), such as the DAM district previously mentioned within Milan's public-procurement initiative. These districts are formal agreements among farmers who share a common vision, production typology, or farming practice, and who coordinate efforts to enhance competitiveness, political visibility, and collective branding strategies. Lombardy has had a regional law[4] regulating agricultural districts since 2007, revised in 2019. Although rooted in national strategies for agricultural modernization and inspired by industrial district models (Bini & Zanolin, 2020), most districts emerged a decade later as part of broader rural development policies.

Currently, Lombardy counts 19 agricultural districts, 5 of which are located within Parco Agricolo Sud Milano. Two districts were formed through strategic partnerships with large retailers, such as Riso e Rane with Esselunga supermarkets and Dinamo with Coop supermarkets. In 2015, within the framework of Expo related initiatives led by Milan's Food Policy Office, the first five districts signed a Territorial Development

[4] It refers to the regional laws D.g.r.n.8/10085 of August 7, 2009, and the D.g.r. 31/07/2019 - n. XI/2040.

Framework Agreement (*Accordo Quadro di Sviluppo Territoriale* - AQST) for the Milan Rural Metropolis project.[5] This agreement aimed to coordinate district activities and integrate them into broader governance strategies led by the Lombardy Region, the Metropolitan City of Milan, and the Municipality of Milan. Through the AQST, the districts share objectives such as enhancing green and blue infrastructures, strengthening and safeguarding agricultural activities, and preserving the complexity of the agrarian landscape as both food-producing and ecologically valuable territory under constant pressure. Two river management authorities, the Fiume Olona and Est Ticino Villoresi consortia, are also signatories to the agreement, echoing some of the aspects discussed in Chap. 5 regarding the management of natural resources.

Beyond production-oriented districts, other models such as bio-districts have emerged, structured specifically around sustainable farming practices and community-building (Basile, 2017). The first bio-district, established in 2004 in Cilento, in the Campania Region, inspired the creation of more than 60 similar initiatives across Italy and other European countries. Lombardy currently hosts four recognized bio-districts—located in Valtellina and Valle Camonica, Bergamo, and the Casalasco Viadanese area—formally acknowledged under Regional Law no. 23/2022. Within the structure of bio-districts, consumer associations play a crucial role in assembling short supply chains and ensuring a stable customer base for farmers. Among these, solidarity purchasing groups (*Gruppi di Acquisto Solidale*—GAS) represent one of the most widespread forms of Alternative Food Networks (AFN) in Italy. GAS are based on agreements between consumers and producers and have contributed to farmers' economic stability, particularly during transitions toward organic certification or product diversification.

GAS and CSA initiatives globally emerged alongside the fair-trade movement of the 1980s and 1990s, rooted in principles of equity and solidarity, and later intersected with local food movements such as Slow Food (Ontano, 2020). In Italy, the first GAS appeared in the 1980s in Parma as Catholic-based purchasing cooperatives[6]. In Lombardy, the earliest GAS

[5] It is a project led by the Metropolitan City of Milan and supported by Lombardy Region through the RURBANCE – Rural Urban Governance EU project 2012–2015 (Regione Lombardia, 2015).

[6] A RAI television reportage from 1980 entitled 'GAS di Parma - Gruppo di acquisto solidale' narrates the story of the first emerging GAS in Italy, specifically in Parma. https://www.youtube.com/watch?v=sNj9q2sswR8 Last accessed on January 16, 2025.

were established between 1993 and 1997, with a significant expansion between 2007 and 2011. According to the CORES Observatory of the University of Bergamo, Lombardy hosted approximately 450 GAS in 2013, representing 25% of the national total. Although dated, this study remains one of the most comprehensive, revealing that most GAS are small-scale, highly participatory, and motivated by product quality, support for local farmers, and environmental sustainability. Each GAS varies in history, structure, and scale, and may be initiated by farmers, associations, families, or neighborhood groups; however, all share a strong social dimension.

To exemplify the structure of GAS, the case of Rimaflow Fuorimercato could represent a compelling intersection between AFN and socio-spatial transformation in the Milanese area. Following the closure of an automotive factory in Trezzano sul Naviglio and inspired by Argentina's *fábricas recuperadas* movement, former workers initiated a regeneration project based on solidarity and circular economy principles. Since 2013, Rimaflow has operated as an independent food network supplying dozens of GAS and direct consumers with locally produced, bio-certified food, while explicitly opposing exploitative labor practices. The regenerated factory now hosts a food market, logistics spaces, co-working areas, and venues for associations and events. Particularly relevant is the spatial dimension of initiatives such as GAS and Rimaflow, which integrate into everyday urban life while sustaining alternative rural economies and food networks.

If the Ruhr Metropolis case underlined the strength of design and planning mechanisms, the Milan case highlights the evolving role of local food policies and bottom-up networks, reinforcing some of the issues presented in the initial Barcelona case. For over a decade, Milan has actively explored and implemented initiatives that underscore the importance of local food governance, both within the urban core and through actions addressing some of the most pressing challenges, such as food access for vulnerable populations and the integration of local farmers into public procurement. This case also emphasizes the role of public institutions as catalysts for knowledge exchange among cities, as well as their capacity to engage a wide range of stakeholders and social groups. More broadly, it reveals the *softer* dimension of food systems—one shaped through policy, coordination, and institutional action—and the impact these mechanisms can have in strengthening urban-rural linkages and connections between local and national governance levels.

Although throughout this book the notion of metropolitan food systems has primarily been addressed from the production side of systems, this chapter emphasizes the importance of examining challenges and needs from a consumer and access perspective. This is not to suggest that food access has been neglected in the book, but rather to highlight specific challenges in urban areas, where consumption is concentrated and vulnerable groups face the greatest difficulties.

In all three cases, addressing local production is framed as a transition toward food systems that are not only local, but also socially and environmentally sustainable, and economically accessible. This challenges the perception of local food as a niche product and positions it instead as central to mainstream production (Born & Purcell, 2006). A holistic understanding is required, one that considers the interconnected challenges explored throughout the book, including natural resource management and biodiversity, social cohesion and solidarity, equity and justice, and decentralized governance and territorial management, as they manifest through farmers, farming practices, access, distribution, and consumption. The final chapter will synthesize these dimensions, drawing conclusions to support planning, policymaking, and governance of agrarian territories and metropolitan food production.

References

Basile, S. (2017). *The experience of Bio-districts in Italy | FAO*. FAO. https://www.fao.org/3/a-bt402e.pdf

Bini, V., & Zanolin, G. (2020). *The DAM - Distretto Agricolo Milanese and the Metropolitan system of agricultural districts*. (p. 18). Urbal.

Born, B., & Purcell, M. (2006). Avoiding the local trap: Scale and food systems in planning research. *Journal of Planning Education and Research, 26*(2), 195–207. https://doi.org/10.1177/0739456X06291389

Brunori, G. (2007). Local food and alternative food networks: A communication perspective. *Anthropology of Food*, (S2), Article S2. https://doi.org/10.4000/aof.430

Coppola, A., Crivello, S., & Haupt, W. (2020). Urban Resilience as New Ways of Governing: The Implementation of the 100 Resilient Cities Initiative in Rome and Milan. In A. Balducci, D. Chiffi, & F. Curci (Eds.), *Risk and resilience: Socio-spatial and environmental challenges* (pp. 113–136). Springer International Publishing. https://doi.org/10.1007/978-3-030-56067-6_8

Diputacio Barcelona. (2020). *Pla de gestio i desenvolupament Espai Agrari de la Baixa Tordera*.

Enthoven, L., & Van den Broeck, G. (2021). Local food systems: Reviewing two decades of research. *Agricultural Systems, 193*, 103226. https://doi.org/10.1016/j.agsy.2021.103226

Federici, S. (2009). On capitalism, colonialism, women and food politics. *Politics and Culture* , Issue 2, Special issue on Food (&) Sovereignty.

Fox-Kämper, R., Kirby, C. K., Specht, K., Cohen, N., Ilieva, R., Caputo, S., Schoen, V., Hawes, J. K., Ponizy, L., & Béchet, B. (2023). The role of urban agriculture in food-energy-water nexus policies: Insights from Europe and the U.S. *Landscape and Urban Planning, 239*, 104848. https://doi.org/10.1016/j.landurbplan.2023.104848

Friedmann, H. (2005). From colonialism to green capitalism: Social movements and emergence of food regimes. In F. H. Buttel & P. McMichael (Eds.), *New directions in the sociology of global development* (Vol. 11, pp. 227–264). Emerald Group Publishing Limited. https://doi.org/10.1016/S1057-1922(05)11009-9

Gentili, A., & Zampetti, G. (Eds.). (2021). *Agroecologia Circolare. Dal campo alla tavola. Coltivare biodiversità e innovazione.* Edizioni Ambiente.

Goodman, D., DuPuis, E. M., & Goodman, M. K. (2013). *Alternative food networks: Knowledge, practice, and politics.* Routledge. https://doi.org/10.4324/9780203804520

Holt Giménez, E., & Shattuck, A. (2011). Food crises, food regimes and food movements: Rumblings of reform or tides of transformation? *The Journal of Peasant Studies, 38*(1), 109–144. https://doi.org/10.1080/03066150.2010.538578

Mazzocchi, G., & Marino, D. (2020). Rome, a policy without politics: The participatory process for a metropolitan scale food policy. *International Journal of Environmental Research and Public Health, 17*(2), 479–497. https://doi.org/10.3390/ijerph17020479

Moragues-Faus, A., & Sonnino, R. (2019). Re-assembling sustainable food cities: An exploration of translocal governance and its multiple agencies. *Urban Studies, 56*(4), 778–794. https://doi.org/10.1177/0042098018763038

Ontano, N. (2020). *Gruppi. Acquisto. Solidale Indagine Esplorativa* [Laurea in Gestione dei Sistemi Alimentari e di Qualità della Gastronomia]. Universita di Parma.

Paül, V. (2015). Los parques agrarios en Cataluña. Breve análisis de la contribución de Josep Montasell a su concepción, desarrollo e implantación. In *El Parque Agrario. Una figura de transición hacia nuevos modelos de gobernanza territorial y alimentaria* (pp. 113–142). Heliconia S. Coop. Mad.

Regione Lombardia. (2015). *Accordo Quadro di Sviluppo Territoriale. Milano Metropoli Rurale.* Regione Lombardia.

Salvador, A. J. (2024). Defining a peri-urban agrarian space as a socio-spatial and environmental unit. Lessons from Barcelona's local food system. In F. Calabrò,

L. Madureira, F. C. Morabito, & M. J. Piñeira Mantiñán (Eds.), *Networks, markets & people* (pp. 72–83). Springer Nature. https://doi.org/10.1007/978-3-031-74704-5_8

Salvador, A. J. (2025). Re-programming agrarian urbanism in South Milan. *The peri-urban question. Landscape Research, 50*(8), 1303–1318. https://doi.org/10.1080/01426397.2025.2526497

Susi Botto, I., & Di Vita, S. (2017). *Oltre l'Expo 2015. Tra dimensione ordinaria e straordinaria delle politiche urbane.* Carocci Editore.

Timpe, A. (2017). *Produktive Parks entwerfen. Geschichte und aktuelle Praxis biologischer Produktion in europäischen Parks.* [Doctoral Thesis, RWTH Aachen]. urn:nbn:de:hbz:82-rwth-2017-005402

Timpe, A. (2019). *Designing productive parks: From romanticizing agriculture towards green infrastructure and productive public space: The Ruhr metropolis as a laboratory for productive parks* [Print]. LIT Verlag.

Tornaghi, C., & Dehaene, M. (2020). The prefigurative power of urban political agroecology: Rethinking the urbanisms of agroecological transitions for food system transformation. *Agroecology and Sustainable Food Systems, 44*(5), 594–610. https://doi.org/10.1080/21683565.2019.1680593

CHAPTER 9

Toward Integrated Planning and Governance

Introduction

As proposed by this book, rethinking the role of planning, governance, and landscape management—and the mechanisms that support local food systems—has become a pressing priority within the socio-ecological transition of territories and communities. As shown in Chap. 3, within the selected contexts, planning and design disciplines have engaged with this question for several decades and have developed ad hoc instruments (e.g., agricultural parks or green belts) that attempt to integrate farmland protection with landscape management, supported by complex governance structures. Chapter 3 retraces their origins, outlines current territorial configurations, and exposes their main limitations, highlighting in particular the challenges posed by ongoing transformations in farming practices and their capacity to adapt to socio-economic and environmental change.

Each case reveals distinct pathways that together exemplify a rich and heterogeneous European context struggling with multiple challenges. Some are social in nature, particularly those related to generational renewal and the aging farming population; others stem from economic dynamics such as land grabbing and from dominant food regimes that have favored industrialized and highly specialized production. Across all cases, however, the most significant challenges are—and will continue to be—environmental, especially in relation to climate change. These challenges intersect with farmers' willingness to embark on often uncertain transitions toward

© The Author(s), under exclusive license to Springer Nature Switzerland AG 2026

A. J. Salvador, *Metropolitan Food Systems*, Sustainable Urban Futures, https://doi.org/10.1007/978-3-032-26016-1_9

diversified production and farming models shaped by decades of restructuring. Such models seek not only economic sustainability but also position farms as active agents within their territories.

From this perspective, the cases also reveal the difficulties farmers face in accessing land and securing legal stability in order to commit to these transitions. The book assembles a complex narrative of how bottom-up practices—understood as farming styles—intersect with planning, governance, and landscape management efforts, and vice versa. As demonstrated, the realm of policymaking, particularly through food policies, has gained significant momentum over the past decade, attempting to redefine food systems and rearticulate relationships between urban cores and their hinterlands. Chapter 2 shows that this momentum is the result of longer-term processes of food system *relocalization* and efforts to re-signify the role of agrarian territories, in which farmers and farms have assumed a central role.

The book introduces a series of thematic lenses through which metropolitan food systems are examined, based on on-site research enriched by field visits and interviews. Each thematic, discussed in Chaps. 5, 6, 7, and 8, is illustrated through individual case-study vignettes that present local food systems as complex assemblages of processes, practices, stakeholders, scales, and temporalities. These elements are brought into dialogue through a shared analytical framework, within which the role of planning—and of planners in particular—emerges as both observant and active, intersecting these processes from a perspective attentive to the dialectics of extended urbanization and global dynamics.

Across the three cases, a constellation of networks, actors, and processes actively shapes agrarian territories and food systems. These include emerging models of agricultural land management promoted by large institutional landowners—such as Fondazione Patrimonio Ca' Granda, supra-local governments, and municipalities—as well as by collective organizations, including agricultural districts. Within this context, agency emerges as a key factor. Interviews and site visits proved instrumental in identifying individuals and institutions that have played—and continue to play—an active role in each case, advocating for transitions toward more sustainable farming practices and food systems.

Agrosocial Urbanism as a Key Element for Reinterpreting the Agrarian Landscape

The concept of *agrosocial urbanism* is introduced in this final chapter as the outcome of a narrative that began by framing *agrarian urbanism* and the *agro-urban form* within the tension between the frameworks of "planetary urbanization" and "planetary rural geographies" (Brenner, 2013; He & Zhang, 2022; Schmid & Topalovic, 2023; Wang et al., 2023). The proposed term embodies a complex and relational approach to understanding how these territories are—and can be—produced through food production and agricultural activities. This entails acknowledging different types and scales of food production, consumption, and governance practices, as well as understanding how spaces of production have come into being, the resources they rely on—including natural, financial, and political resources—their historical evolution, current challenges, and possible future trajectories.

Although the term agrarian urbanism has been employed across various scholarly debates—evolving from a perception of the agrarian as a planning ideal to debates that disclose the complexity and entanglements of the agrarian-urban continuum, and more recently to those that rethink the metabolic relations between cities and countrysides (Bathla & Lazzarini, 2025; Fera et al., 2025)—the definition proposed here introduces a non-neutral term: An operative concept that supports a scenario in which the socio-ecological transition of agrarian territories is fundamental for cities and societies. It also seeks to converge the various efforts emerging from different approaches, ranging from the everyday practices of farmers to policymaking and governance, while providing a framework for planners and planning institutions to move beyond mere farmland preservation and actively engage with these processes. This definition aligns with the pathways outlined by Tornaghi and Dehaene (2020) in their concept of *agroecological urbanism* but it is less explicit regarding the production (farming) model itself. Instead, it emphasizes the social processes that sustainable farming practices can also generate, along with other aspects articulated by *agrosocial urbanism*.

The term *agrarian landscape* is used in this section's title and throughout the book as a bridging concept across the many disciplines that engage with landscape in order to address its complexity. Building on the work of Emilio Sereni (1961), the agrarian landscape is understood as the outcome of long-term interactions between agricultural practices, technologies, and cultural values, materially expressed in space and shaped by broader socio-economic structures. In this sense, *agrosocial urbanism* embodies a historical perspective that acknowledges and values the role of agricultural and social practices in shaping territories, while simultaneously emphasizing the need to actively imagine and design their futures within the socio-ecological transition of food systems, places, and communities.

From an epistemological perspective, this perspective suggests that the triad of planning, land management, and governance in agriculture and food production should engage more explicitly with scholarly debates introduced by critical food studies, as well as with perspectives from movements such as *urban political ecology* and *degrowth* (Borras, 2023; Kaika et al., 2023; Savini, 2024), particularly in their efforts to address the metabolic relationships between territories. The urban political ecology lens proves especially valuable for approaching local food systems, as it foregrounds the "co-evolutionary dimensions of social and environmental change" and promotes "an expanded conception of urban metabolism and the circular dynamics of urban space" (Gandy, 2023, p. 61). As an interdisciplinary field, urban political ecology highlights the contested nature of environmental crises and the plurality of interpretations surrounding them, while establishing connections between local, global, and planetary processes. It enables an understanding of how planetary crises and extended urbanization impact local territories, and encourages a rethinking of issues such as limits to growth, resource scarcity, demographic change, and social justice within alternative frameworks to dominant capitalist models (Leff, 2013; Torre, 2023), particularly in response to agricultural and food regimes that have historically decoupled production from natural processes while tightly linking it to market rationalities (McMichael, 2014).

This research project has generated substantial knowledge on the multiple elements that shape agrarian territories socially and spatially, while also proposing a framework through which planning and governance can engage with different levels, scales, and mechanisms. In this context, the role of politics and intentionality emerges as a crucial resource, particularly when grounded in robust knowledge. For this reason, urban and territorial studies scholars have a key responsibility to produce situated and tangible knowledge of agrarian territories and of the spatial and environmental dynamics embedded in food production systems. The emergence of *degrowth* as a potential pathway toward an equitable downscaling of production and consumption offers a particularly relevant perspective, as it points to specific spatial relationships that foreground the social and environmental carrying capacities of territories (Savini, 2024). This approach stands in contrast to localist movements that emerged in the early 2000s, which often focused on rebranding consumption patterns while downplaying broader social struggles, and were later absorbed into new forms of "green capitalism" (Friedmann, 2005).

In re-establishing spatial relationships and redefining the scope and extent of local food systems, urban and territorial studies—together with planning disciplines—can play a central role. As discussed in Chap. 1, this challenge requires rethinking the relationship between cities and their immediate hinterlands (Brenner & Katsikis, 2020; Schmid & Topalovic, 2023), while also exploring alternative territorial models that prioritize forms of economic development capable of recognizing and preserving social and environmental assets. One such model is the *urban bioregion* (Atkinson, 1992; Fanfani & Matarán Ruiz, 2020), which was introduced in Chap. 3 as a spatial framework for rearticulating production, governance, and territorial identity.

To bring coherence to these multiple levels of analysis, the concept of *agrosocial urbanism* is introduced as both a synthetic definition and a conceptual framework capable of framing and orienting the overall research process. It proposes a relational approach to understanding (peri-urban) agrarian landscapes through the intertwined dynamics of food production, governance, and socio-ecological transformation. Within this framework, agriculture is understood not merely as a productive activity, but as a territorial practice that connects actors, resources, and urban-rural processes. Its core principles recognize the plurality of scales involved in food production and governance, emphasize the role of agricultural and social practices in shaping territories, and examine how agrarian spaces have been historically constructed through the mobilization of natural, financial, and political resources. In doing so, *agrosocial urbanism* situates agrarian futures within a broader horizon of socio-ecological transition and orients planning toward emerging and unresolved challenges.

Metropolitan Food Systems and Integrated Planning and Governance Processes

As evidenced across all cases, the geography of food production is rich and complex, even when examined within a specific territorial context. The term *local* (food systems) is intentionally replaced, as this research demonstrates that it carries multiple meanings, scopes, and dimensions—particularly when approached from a perspective of capitalist accumulation. Instead, emphasis is placed on understanding the geography of food production through an awareness of the relational and variable scales of the multiple systems embedded within it. This perspective is particularly relevant for spatial planning and strategic frameworks, especially in relation to how they intersect with agriculture and food production.

The scale of analysis adopted across the three contexts was therefore not incidental. The initial inquiry focused on the *agricultural park* as a specific planning instrument, with the intention of reassessing its role under contemporary conditions. However, early findings revealed a relative lack of scholarly research and limited engagement by planning professionals and institutions with agriculture and food production from a broader territorial perspective. This became especially evident when examining the polycentric structure of the Ruhr Metropolis and the political challenges faced by the Metropolitan City of Milan in governing the 47,000 hectares of *Parco Agricolo Sud*, which spans 61 municipalities. This shift in scale enabled a more explicitly territorial interpretation of agricultural practices and of the differentiated challenges faced by specific production areas.

The Barcelona case offers perhaps the most tangible example of how the geography of food production is structured through identifiable peri-urban agrarian spaces, each characterized by specific production orientations as well as distinct challenges and vulnerabilities. This perspective also prompted further inquiry into related issues, such as water provision and green infrastructure. As discussed throughout the book, city-regions and metropolitan areas have become a central focus in urban and territorial planning research, particularly with respect to demographic, economic, mobility, governance, and morphological dynamics (Balducci et al., 2019). While relevant work has explored the integration of food into the metropolitan agenda (e.g., Morgan, 2009; Wascher et al., 2015), such research has largely concentrated on assessing food self-provisioning capacities.

By contrast, this book has focused on developing a comprehensive spatial, social, and environmental assemblage of how metropolitan territories engage with agriculture and food production. This includes an analysis of planning approaches, the social value attributed to agrarian spaces, and the challenges these territories face in functioning simultaneously as environmental assets and as spaces for food provisioning.

Concerning the tools available for food planning, governance, and landscape management, Chapter 3 extensively discussed the agricultural park and similar types of parks (both landscape and productive) as mechanisms conceived to preserve and revitalize agricultural production areas (Fanfani, 2019; Ferraresi & Rossi, 1993). Despite not being formally recognized as a statutory planning instrument and exhibiting a low rate of implementation—as evidenced by Zazo Moratalla and Paul (2022) in the Spanish context—it remains a distinctive tool capable of capturing the territorial complexity of agrarian communities in their social and economic dimensions, while seeking to transform them into environmental and productive assets.

The Baix Llobregat case stands out as a rare and difficult-to-replicate example for many of the reasons discussed throughout the book. Nevertheless, it has endured for more than 25 years, demonstrating an ability to adapt to the evolving challenges faced by farmers and agrarian territories, while responding to societal demands and political pressures. A pending challenge, however, concerns how the park—as both an institution and a collective of farmers—positions itself within the socio-ecological transition expected for agricultural and food production systems. The agricultural park, as a planning and governance mechanism for peri-urban agrarian territories, thus remains relevant and deserves renewed consideration, particularly in light of current global challenges.

Table 9.1 attempts to synthesize and place in dialogue the spheres of planning and governance with those of agriculture and food production, outlining the different levels, tools, mechanisms, and forms of interaction among them, as well as the range of possible stakeholders involved. The table also reinforces the idea that some strategies emerge from institutional processes operating at different scales, while others originate from bottom-up initiatives and practices that nonetheless play a significant role in shaping the concept of *agrosocial urbanism*.

In addition, Table 9.1 incorporates dimensions such as resource management and territorial governance which are primarily addressed through larger-scale spatial planning and green infrastructure strategies. These strategies are sometimes conceived for specific elements, such as rivers or systems of green spaces, and more recently for climate adaptation and mitigation strategies purposes. Examples include the regional parks system and river contracts in Lombardy, as well as spatial planning strategies aimed at revitalizing critical natural elements within the Regionalverband Ruhr. Landscape and agricultural parks likewise exemplify territorial

Table 9.1 Tools and strategies for food planning and governance and farmland management

Acting level	*Examples of tools and strategies*	*Stakeholders*
Territorial strategies	Water, climate change adaptation, and mitigation strategies	National governments and supra-national and international institutions
Regional strategies	Spatial planning, green infrastructure and landscape management strategies, and management entities (e.g., river management authorities)	Regional and metropolitan levels of government and other public institutions working at this scale
Local and supra-local planning	Landscape parks, agricultural and productive parks, management entities (e.g., park management authorities) and public land management models	Local governments and agreements between municipalities, constituted metropolitan areas
Policies and strategies	Food policies and other documents, charters, development and management plans	Local and regional governments, farmers and consumers associations
Programs and place-based policies	Land banks, farmer's markets, public procurement, long-term lease contracts, marketing and communication strategies, educational programs	Local governments, landowners, park management entities
Networks	Consumer groups (CSA, SoLaWI, GAS) and alternative food networks, farmers groups and associations, production districts, and landowners	Farmers, consumers, associations, and landowners
Farming models	Food production, other services produced by farming, and non-farming services, incentives and subsidies	Farmers and consumers, associations, local, regional, and national governments

Source: Elaborated by the author

governance approaches and efforts to preserve farmland and farming systems.

Across all three cases, there are significant examples of how planning and design engage with agricultural land use in more complex ways and across multiple scales. This perspective also foregrounds policymaking as a key resource for producing frameworks—such as policy documents or charters—that introduce food and production issues into political agendas and facilitate collaboration among diverse actors, including farmers, associations, and consumer groups. These mechanisms and models have further enabled the development of programmatic and place-based policies tailored to the specific challenges faced by farmers and agrarian territories.

The Baix Llobregat agricultural park in Barcelona is distinctive in its capacity to integrate planning and management through a dedicated institutional entity—the park's consortium—whose role has been continuously adapted to the needs of this specific territory. Programs such as farmers' markets, educational initiatives, the land bank, and targeted marketing strategies constitute ad hoc mechanisms that support farmers and contribute to building a recognizable territorial identity.

A relevant contribution by this research lies in its inquiry into the role of a wide range of agriculture- and food-related networks and the actors emerging from these relationships. This dimension has received comparatively limited attention in planning and territorial studies, yet it plays a significant role in shaping strategies for agrarian territories and in envisioning their futures. Some of these networks—such as consumer groups—are directly integrated into farming models, while others, including farmers' groups and associations, have historically contributed to social cohesion within agrarian areas. At the same time, agricultural districts (*distretti agricoli*) represent more recent forms of collaboration among farmers. Although initially driven by economic motivations linked to competitiveness within corporate food production models, these districts are increasingly exploring collective pathways toward transition, including shifts to organic farming practices or the promotion of specific production areas, as exemplified by the bio-districts in Lombardy.

This research also examined the role of large landowners, particularly in the cases of the Ruhr Metropolis and Milan. In the latter, the active engagement of a major institutional landowner committed to farming and local food production—such as Fondazione Patrimonio Ca' Granda—has revealed the potential and impact of this type of actor. Several of the foundation's programs are comparable to the management practices of the Baix Llobregat Agricultural Park consortium and, in some instances, are even more ambitious, including initiatives aimed at enabling short supply chains and systematically monitoring soil and water health. Other relevant models discussed include the Regionalverband Ruhr's newly drafted farmland management guidelines, the City of Bochum's compensation strategies, and the Municipality of Milan's renewed efforts to influence agricultural practices on publicly owned land. In these cases, contractual conditions—particularly their duration—emerge as decisive factors in steering farming toward models that integrate social and environmental objectives.

These observations foreground agriculture not only as an economic activity but also position the farm holding as a key unit in transforming food systems and mediating their relationship with planning and governance frameworks. This research has fully acknowledged the role of the individual farm as a catalyst for collective change. While recent scholarship in urban and territorial studies has focused on developing typologies of urban and peri-urban agricultural practices (e.g., Jansma et al., 2024; Lohrberg et al., 2015), this work advances the field by embedding farm-level agricultural practices within a broader analytical framework that connects food production with planning, governance, and land management. As demonstrated throughout the study, the tools, scales, actors, and mechanisms through which agriculture and food systems intersect with territorial planning and governance are highly diverse.

Taken together, the three cases examined in this book reveal the multiplicity of actors and dimensions involved in shaping agrarian landscapes and food systems. While this research does not aim to propose a specific methodological framework, it provides empirical evidence of the complexity and richness inherent in territories devoted to agricultural activity and food production. It argues for approaching the challenge of feeding cities by embracing this complexity and by advancing integrated planning and governance processes capable of addressing the socio-ecological transition of food systems, places, and communities by strengthening the relationships between society and nature, and between production and consumption.

References

Atkinson, A. (1992). The urban bioregion as "sustainable development" paradigm. *Third World Planning Review, 14*(4), 327–355. https://doi.org/10.3828/twpr.14.4.q1086507jk68vjvh

Balducci, A., Fedeli, V., & Curci, F. (Eds.). (2019). *Post-metropolitan territories: Looking for a new urbanity.* Routledge. https://doi.org/10.4324/9781315625300

Bathla, N., & Lazzarini, L. (2025). Towards an agrarian urbanism beyond the rural–urban divide. *Landscape Research, 50*(8), 1261–1271. https://doi.org/10.1080/01426397.2025.2591265

Borras, S. M. (2023). Politically engaged, pluralist and internationalist: Critical agrarian studies today. *The Journal of Peasant Studies, 50*(2), 449–489. https://doi.org/10.1080/03066150.2022.2163164

Brenner, N. (Ed.). (2013). *Implosions/explosions. Towards a study of planetary urbanization*. JOVIS. https://doi.org/10.1515/9783868598933

Brenner, N., & Katsikis, N. (2020). Operational landscapes: Hinterlands of the Capitalocene. *Architectural Design, 90*(1), 22–31. https://doi.org/10.1002/ad.2521

Fanfani, D. (2019). Agricultural park in Europe as tool for agri-urban policies and design: A critical overview. In E. Gottero (Ed.), *Agrourbanism: Tools for governance and planning of agrarian landscape* (pp. 149–169). Springer International Publishing. https://doi.org/10.1007/978-3-319-95576-6_10

Fanfani, D., & Matarán Ruiz, A. (2020). Introduction to bioregional planning. Relocalizing cities and communities for a post-oil civilization. In D. Fanfani & A. Matarán Ruiz (Eds.), *Bioregional planning and design: Volume I: Perspectives on a transitional century* (pp. 1–15). Springer International Publishing. https://doi.org/10.1007/978-3-030-45870-6_1

Fera, A., Martinelli, A., & Salvador, A. J. (2025). Of visions and practice: Interpreting contemporary forms of agrarian urbanism. *Landscape Research, 50*(8), 1371–1385. https://doi.org/10.1080/01426397.2025.2563723

Ferraresi, G., & Rossi, A. (Eds.). (1993). *Il parco come cura e coltura del territorio. Un percorso di ricerca sull'ipotesi del parco agricolo*. Grafo edizioni.

Friedmann, H. (2005). From colonialism to green capitalism: Social movements and emergence of food regimes. In F. H. Buttel & P. McMichael (Eds.), *New directions in the sociology of global development* (Vol. 11, pp. 227–264). Emerald Group Publishing Limited. https://doi.org/10.1016/S1057-1922(05)11009-9

Gandy, M. (2023). Urban political ecology, versus ecological urbanism. In *Turning up the heat. Urban political ecology for a climate emergency* (pp. 56–66). Manchester University Press.

He, S., & Zhang, Y. (2022). Reconceptualising the rural through planetary thinking: A field experiment of sustainable approaches to rural revitalisation in China. *Journal of Rural Studies, 96*, 42–52. https://doi.org/10.1016/j.jrurstud.2022.10.008

Jansma, J. E., Veen, E. J., & Müller, D. (2024). Beyond urban farm and community garden, a new typology of urban and peri-urban agriculture in Europe. *Urban Agriculture & Regional Food Systems, 9*(1), e20056. https://doi.org/10.1002/uar2.20056

Kaika, M., Keil, R., Mandler, T., & Tzaninis, Y. (2023). *Turning up the heat. Urban political ecology for a climate emergency*. Manchester University Press.

Leff, E. (2013). Ecologia Política: Uma perspectiva latino-americana. *Desenvolvimento e Meio Ambiente, 27*. https://doi.org/10.5380/dma.v27i0.32510

Lohrberg, F., Licka, L., Scazzosi, L., & Timpe, A. (Eds.). (2015). *Urban agriculture Europe*. Jovis.

McMichael, P. (2014). *Food regimes and agrarian questions* (pp. 1–196). Practical Action Publishing. https://doi.org/10.3362/9781780448787
Morgan, K. (2009). Feeding the city: The challenge of urban food planning. *International Planning Studies, 14*(4), 341–348. https://doi.org/10.1080/13563471003642852
Savini, F. (2024). Strategic planning for degrowth: What, who, how. In *Planning theory*. Sage UK. https://doi.org/10.1177/14730952241258693
Schmid, C., & Topalovic, M. (Eds.). (2023). *Extended urbanisation. Tracing planetary struggles*. Birkhäuser.
Sereni, E. (1961). *Storia del paesaggio agrario italiano* (1st 2020). Editori Laterza.
Tornaghi, C., & Dehaene, M. (2020). The prefigurative power of urban political agroecology: Rethinking the urbanisms of agroecological transitions for food system transformation. *Agroecology and Sustainable Food Systems, 44*(5), 594–610. https://doi.org/10.1080/21683565.2019.1680593
Torre, S. (2023). Che cos'è l'ecologia politica? In L. Pellizzoni (Ed.), *Introduzione all'ecologia politica* (pp. 29–56). Il Mulino.
Wang, C.-M., Maye, D., & Woods, M. (2023). Planetary rural geographies. *Dialogues in Human Geography*, 20438206231191731. https://doi.org/10.1177/20438206231191731
Wascher, D., Zasada, I., & Sali, G. (2015). Tools for metropolitan food planning—A new view on the food security of cities. In *The governance of city food systems* (pp. 68–97). Fondazione Giangiacomo Feltrinelli.
Zazo Moratalla, A., & Paul, V. (2022). What is an agricultural park? Observations from the Spanish experience | Elsevier enhanced reader. *Land Use Policy, 112*. https://doi.org/10.1016/j.landusepol.2021.105584

Index[1]

[1] Note: Page numbers followed by 'n' refer to notes.

© The Editor(s) (if applicable) and The Author(s), under exclusive license to Springer Nature Switzerland AG 2026
A. J. Salvador, *Metropolitan Food Systems*, Sustainable Urban Futures, https://doi.org/10.1007/978-3-032-26016-1

P

R

S

T

U

W

The manufacturer's authorised representative in the EU is Springer Nature Customer Service Centre GmbH, Europaplatz 3, 69115 Heidelberg, Germany. If you have any concerns regarding our products, please contact ProductSafety@springernature.com

Printed and bound by CPI Group (UK) Ltd, Croydon, CR0 4YY

08/10/2026

02230045-0001